ISBN-13: 978-1540887481

ISBN-10: 1540887480

SCSC Publication Number: SCSC-127B

The SCSC is the professional network for sharing knowledge about safety-critical systems. It brings together: engineers and specialists from a range of disciplines working on safety-critical systems in a wide variety of industries; academics researching the arena of safety-critical systems; providers of the tools and services that are needed to develop the systems; and the regulators who oversee safety. Through publications, seminars, workshops, tutorials, a web site and, most importantly, at the annual Safety-critical Systems Symposium (SSS), it provides opportunities for these people to network and benefit from each other's experience in working hard at the accidents that don't happen. It focuses on current and emerging practices in safety engineering, software engineering, and product and process safety standards.

This document was written by the Data Safety Initiative Working Group (DSIWG), which is convened under the auspices of the SCSC. The document supports the DSIWG's vision, which is to have clear guidance that reflects emerging best practice on how data (as distinct from software and hardware) should be managed in a safety-related context. This update takes account of the consensus that a process-based guidance document will complement existing safety management processes, making it more useable. It was formally released at SSS'17, 7-9 February 2017.

Comments on this document are actively encouraged. These can be emailed to: comments@data-safety.scsc.uk. Alternatively, a comments submission form is available at: data-safety.scsc.uk/comments.

# Data Safety Guidance

## The Data Safety Initiative
## Working Group [DSIWG]

January 2017

# Change History

| Version | By | Status | Date |
|---------|----|--------|------|
| 1.0 | The DSIWG Team | First draft for external review | 31-JAN-2014 |
| 1.1 | The DSIWG Team | (Internal edition for DSIWG use only) | 09-DEC-2014 |
| 1.2 | The DSIWG Team | For publication at SSS'15 | 23-JAN-2015 |
| 1.3 | The DSIWG Team | For publication at SSS'16 | 29-JAN-2016 |
| 2.0 | The DSIWG Team | For publication at SSS'17 | 30-JAN-2017 |

## Changes Since the Last Edition

The most significant changes to this edition of Data Safety Guidance have been made in response to feedback on the previous edition. In particular, the aim has been to move towards a document that is less theoretical and more practical; that is, a document that can more readily be used by the intended readership, which includes safety practitioners, auditors, regulators and managers.

The most obvious changes are a more explicit focus on a Data Safety Management Process, which includes a number of objectives. Data safety is still a comparatively young discipline and, as such, it was inappropriate to provide a long list of detailed objectives. Hence, the identified objectives are deliberately small in number and high-level in nature.

Another part of making the guidance more practical was to ensure it could be applied in a wide range of situations. To support this, the new edition contains advice on tailoring.

A host of smaller changes have also been introduced. Examples include: an updated collection of historical accidents and incidents; a new supplier data maturity questionnaire; a discussion on how Data Safety Assurance Levels relate to Safety Integrity Levels (or equivalent); and a comment on the similarities and differences between data and software.

Finally, this edition has been granted a Safety Critical Systems Club (SCSC) publication number. This has been adopted to allow the SCSC to better manage its portfolio of publications. It also has the advantage of allowing this document to be referenced in a very brief style, namely: SCSC-127B.

# Foreword

**Data is here. Data is growing. Data is causing harm.**

**Data is here** The way that systems are being designed and built is changing. Whereas data was once used simply to configure (and leave configured) a system, its use is rapidly expanding as developers seek to exploit the flexibility provided by data-driven systems. Consider, for example, the importance of data defining the layout of Britain's railway signals, data which indicates the position of underwater obstructions in nautical channels or data that records a patient's treatment history. Organisations now make significant decisions (including safety-related decisions) based solely on data held in systems. Hence, organisations need to safely manage, control and process their data. In particular, key Data Properties that preserve safety must be actively managed.

**Data is growing** There are at least two reasons why the use of data has grown and, equally important, why it is expected to continue to grow. The first relates to the rapid expansion of the area loosely termed "Big Data". The second is the growing use of systems of systems, where data is the lifeblood that connects together disparate elements and allows a cohesive capability to be built. Put simply, the need to address data-related issues is a pressing problem and will continue to be so.

**Data is causing harm** Strictly speaking, this is not accurate; by itself data can neither cause nor prevent harm. However, mistakes introduced in data, or the inappropriate use of data, within safety-related systems have been factors in a number of documented accidents and incidents. Examples include: aircraft attempting to take-off from the wrong runway (and consequently crashing); ships running aground; and patients being exposed to higher than planned doses of radiation.

Against this background, the Data Safety Initiative Working Group (DSIWG) was established under the auspices of the Safety Critical Systems Club. The DSIWG's aim is to develop clear, cross-sector guidance that reflects emerging best practice on how data (as opposed to software or hardware) should be managed in a safety-related context. For the most part, this guidance is based on well-established techniques. What is new, however, is the explicit and relentless focus on data, making it a "first class citizen" within system safety analyses. By doing so, this guidance should help organisations identify, analyse, evaluate and treat data-related risks, thus reducing the likelihood of data-related issues causing harm in the future.

# Quick Start Guide

> *Data really powers everything that we do.*
> **Jeff Weiner**

- Systems are changing. The role of data is becoming more prominent. Hence, data needs to be considered as a "first class citizen" in system safety analyses. This will help mitigate organisational and system level risks associated with the use of data.

- A Data Safety Management Process (Section 2) has been developed. This includes:

  - Assessments to help establish the appropriate context (Section 3);

  - Methods for identifying risks, including top-down and bottom-up approaches (Section 4);

  - An approach for analysing risks, which is based on Assurance Levels (Section 5); and

  - Support for the evaluation and treatment of risks (Section 6).

- A worked example is provided (Section 7).

- A collection of appendices provide more detail, including:

  - A brief summary of previous accidents and incidents in which data was potentially a causal factor (Appendix A);

  - An Organisation Data Risk assessment questionnaire (Appendix B);

  - A Data Safety Culture questionnaire (Appendix C);

  - A questionnaire to help assess "data maturity" of a supplier (Appendix D);

  - A list of Data Types (Appendix E);

  - A collection of Hazard and Operability Study Guidewords (Appendix F);

  - The suggested contents of a Data Safety Management Plan (Appendix G);

  - Lists of acronyms, definitions and glossary entries (Appendix H); and

  - A collection of references (Appendix I).

# Contents

# 1  Introduction

*We're entering a new world in which data may be more important than software.*
**Tim O'Reilly**

## 1.1  Motivation

On 4 March 2015 a Turkish Airlines A330-303 was arriving at Kathmandu-Tribhuvan Airport (KTM), Nepal. An initial approach was abandoned. On the second approach the aircraft touched down to the left of the runway centre line with the left hand main gear off the paved surface. It ran onto soft soil and the nose landing gear collapsed. Following the accident the aircraft was written off. The displaced landing occurred because the Flight Management Guidance System (FMGS) navigation database contained inaccurate and out-of-date coordinates. This incident is one of a number where inappropriate data has featured as a causal factor. Several other examples are presented in Appendix A.

Through incidents like the one outlined above, it has become increasingly clear that system safety depends not only on the system's hardware and software, but also on the data it accepts, generates, processes and produces. As a result, the Safety Critical Systems Club (SCSC) has promoted efforts to investigate the issue of data in safety-related systems. (The term "safety-related systems" is used throughout this document. It includes systems where failure will lead to immediate harm, as well as systems where failure may not lead to immediate harm but could contribute to its likelihood of occurring.)

Safety-related data takes many forms; examples include:

- Data used by an application, e.g., patient medical records in a hospital;

- Data about a system itself, e.g., configuration data for a satellite navigation system; and

- Data about users of a system, e.g., operator competence data in a nuclear power plant.

One reason for the growing importance of data is a change in the way that systems are designed and built. Traditionally, components were engineered to specific standards and then configured by data to perform a bespoke role. Today, safety-related systems are becoming more data-intensive and data-centric. A key feature of this type of system is that the criticality is inherently with the data, rather than being in a directly controlling function. These data-intensive, safety-related systems are often used as decision support or advisory systems, which support a trained and experienced operator, who may be able to detect and correct data problems. However, data is now so complex and of such a large volume it is increasingly unlikely that a user would spot the data errors, and it would be unreasonable to expect them to do so.

There are also industry trends which make a data safety initiative very timely: the drive towards "Big Data" systems means that safety-related data is being used in more and varied ways, and as part of very large aggregated databases, often via the Internet. Increasing use of Systems of Systems (SoS) technology means that data systems are becoming more highly connected using data from a variety of sources. Hence, the mapping and translation of data between diverse systems becomes more important and more challenging.

Despite the ever growing importance of data, current safety standards and regulations focus strongly on the development of systems, hardware and software. Data-related aspects are covered comparatively poorly, if they are covered at all. In most sectors, data as separate entity has hardly been considered (the aviation domain is a notable exception). Sometimes certain types of safety-related data have been identified, but then little guidance is provided on how to determine or manage the associated risks. In many standards, data is treated in a similar way to software, although the properties it exhibits can be very different. There are also several sectors which now produce and manipulate safety-related data in vast quantities, and which are not thought to be covered by any existing safety standards; the Police and Criminal Justice sector is one such example.

The significant and continually growing importance of data, combined with the apparent lack of guidance on how associated risks can be managed, provide the background to the SCSC's data-safety initiative. The main motivation of this document is to begin the process of raising the prominence of data to that of a "first class citizen" in the design, implementation and operational use of safety-related systems.

## 1.2    Aim and Scope

This guidance document aims to:

- Describe the data safety problem;

- Provide methods for identifying and analysing levels of risk; and

- Recommend methods and approaches for evaluating and treating those risks.

It should be noted that, whilst they are considered mature enough to be useful, the contents of the document represent current thoughts on what is a complex and evolving area. Furthermore, in order to allow it to be produced within a reasonable timescale, this edition focuses on key items; it is not intended to be exhaustive.

This document has been written for a wide readership. Its target audience covers all those who have an interest in, or a responsibility for, safety-related data within systems, including: Managers, Developers, Safety Engineers, Assurers (including Independent Safety Auditors), Regulators and Operators.

The breadth of readership is also intended to cover a number of different sectors. As such, the document identifies a wide spectrum of safety-related data that exists in many forms within systems: from specification and requirements data, to maintenance and disposal data, and everything in between. In particular, this document is not just concerned with numerical or well-structured data used during system operation.

The document is not intended to replace or supersede any existing material, whether that be sector-specific or of a more general nature. Nor does it (currently) provide an acceptable means of compliance to any particular standard. Instead, the intent is for this guidance to be used as a supplement to current standards. In the longer-term the hope is that future standards documents will take up relevant concepts, approaches and methods from those described here. It is also hoped that organisations will encapsulate some of these concepts, approaches and methods in their own safety management processes.

## 1.3    Continuous Improvement

Like the continual improvement processes inherent in all best practice safety management systems, the Data Safety Guidance (DSG) is also subject to continual improvement efforts, which directed by the DSIWG. The DSIWG is focussing on a number of improvement areas for the next issue of this guidance, including attempting to integrate the DSG with existing safety standards. Since the DSG is purposefully industry-neutral, the list of standards and other documentation of interest is extensive. Key examples include Defence Standards 00-055 and 00-056, the guidance provided by NHS Digital, and software-related standards such as IEC 61508.

The DSIWG welcomes both new members and requests for particular information or focus areas within this continual improvement activity.

## 1.4    Data Safety Assurance Principles

Hawkins *et. al.* established some generic software safety assurance principles, which are commonly referred to as "4 + 1" [17]. Given the close links between software and data it is helpful to consider these principles from a data-safety assurance perspective. The results are detailed below, with each principle being considered in turn.

### 1.4.1    Principle 1: Data safety requirements shall be defined to address the data contribution to system hazards

Data pervades active system operation, as well as the system's specification, realisation, verification, validation, certification, maintenance, and retirement. Moreover, data may be passed from one system to another; sometimes over a significant period of time. Data may be assimilated, and converted from prior uses into new uses, or simply used as is by many systems. It is stored in media whose storage integrity decays. The system context for data safety requirements may be specific to a particular system's (or [safety] engineering process's) use of the data, or it may be generalised to a class of related systems. Hence data safety requirements are needed for any safety-related system that interacts with data.

### 1.4.2    Principle 2: The intent of the data safety requirements shall be maintained throughout requirements decomposition

Data safety requirements establish the system's safety properties for data, for the system's use of data, for the management of data and for the engineering lifecycle of both the system and its associated data. The system's requirements hierarchy must preserve the intent of the data safety requirements (and hence the system's safety-related Data Properties). Moreover, the applied engineering process for both the system's realisation and subsequent lifecycle stages shall demonstrate that the data safety properties are preserved.

### 1.4.3    Principle 3: Data safety requirements shall be satisfied

Evidence is required that the system satisfies all of the data safety requirements imposed on it for all anticipated operating conditions. Moreover, the data safety requirements that pertain to the data's lifecycle outside of the system shall be evidentially demonstrated prior to the system acting on such data, or else the system shall be able to adequately defend against unmet data safety requirements. In other words, either the data can be shown to demonstrate the required Data Properties prior to being used, or the system can implement adequate defences and mitigations against data that does not conform to the required safety properties.

### 1.4.4    Principle 4: Hazardous system behaviour arising from the system's use of data shall be identified and mitigated

This is an intentionally broad statement because data is conceptual and not physical; it is the contextualised use of data that could result in a system hazard. Data Principle 1 deals with system level hazards arising from data, whereas Data Principle 4 is concerned with hazards that arise from the way the system uses its data; that is, whether the system's design and implementation introduce further hazards. An example is a ship navigation system's display of hydrographic chart data, where a wide field display results in small shallow underwater features disappearing (due to image scale) when it is critical that situational awareness of such hazards is maintained.

### 1.4.5 Principle 4+1: The confidence established in addressing the Data Safety Assurance Principles shall be commensurate to the contribution of the data to system risk

The confidence in the evidence that demonstrates establishment of the first four Data Safety Assurance Principles shall be proportionate to the contribution data has with the system hazards.

# 2 Data Safety Management Process

> *Errors using inadequate data are much less than those using no data at all.*
> **Charles Babbage**

## 2.1 Overview

The Data Safety Management Process has been structured around ISO 31000 [20], taking into account the Data Safety Assurance Principles. This means that a relatively simple process, which focuses on data-related aspects, can be presented here and applied by individual organisations. If appropriate, organisations can also integrate this data-specific process into their own safety management processes and, in due course, their Safety Management System.

To simplify the presentation, the process is presented as a series of sequential phases. In practice, a degree of iteration is likely to be required (e.g., measures adopted to treat risks may lead to refined system design and revised risk analyses). Likewise, it may be appropriate for some parts of the process to run in parallel (i.e., a subsequent phase may start before a preceding phase has finished).

A key aim is that the Data Safety Guidance can be applied across all industry sectors and to a wide range of system types. This extensive coverage can only be achieved through tailoring: what is required for large-scale, safety-critical systems would be excessive for small-scale systems that pose significantly less risk. Suggestions for tailoring are included in this document. Alternative levels of tailoring may be adopted, if sufficient justification is provided.

Although the Data Safety Management Process is based on the structure of ISO 31000, there are some differences between the two. These include:

- The ISO standard's "Establish Context" phase is concerned with an organisation adopting a risk management system; within this guidance document that phase is extended to apply to specific projects;

- The ISO standard considers risk as being synonymous with uncertainty in outcome (i.e., some risks may be beneficial and hence it may be desirable to take actions to increase their likelihood); within this guidance document all risks are considered to have adverse effects; and

- The ISO standard separates the "Evaluate Risks" and "Treat Risks" activities; within this guidance document it has been convenient to combine these into a single phase.

ISO 31000 also recommends two activities that run in parallel with risk assessment. These activities: (1) monitor and review the risk assessment process; and (2) communicate and consult with the stakeholders about the risk assessment process. Aspects like "monitor", "review", "communicate" and "consult" are taken to be part of normal project activities; items like the Organisational Data Risk (ODR) assessment and the Data Safety Management Plan (DSMP) are intended to assist from the perspective of data safety.

## 2.2   Summary Figure

| | Objectives | Activities | Main Guidance Material | Outputs |
|---|---|---|---|---|
| **Establish Context** | Key stakeholders and necessary approvers are identified<br><br>Interfaces are defined and controlled<br><br>The Data Safety assessment is planned<br><br>Data Artefacts are identified | Describe the organisational context<br><br>Describe the system context<br><br>Plan the assessment<br><br>Identify Data Artefacts | Organisational Data Risk (ODR) assessment<br><br>Data Safety Culture questionnaire<br><br>System lifecycle, producers and consumers<br><br>Supplier data maturity questionnaire<br><br>Data Types | A list of key stakeholders and necessary approvers for data safety activities<br><br>An interface control plan or list of control measures<br><br>An estimate of the level of data-related risk (e.g., an ODR rating)<br><br>A plan for the remaining parts of the data safety assessment<br><br>A collection of Data Artefacts |
| **Identify Risks** | Generic and historic examples of data-related issues are reviewed<br><br>Risks are identified and linked to Data Artefacts and Data Properties | Review the general, historical perspective<br><br>Conduct a top-down approach<br><br>Conduct a bottom-up approach<br><br>Update planning documents | Historical accidents and incidents<br><br>Generic ways data can cause problems<br><br>Data Properties<br><br>Hazard and Operability Study Guidewords | A description of the process used for risk identification<br><br>List of risks linked to Data Artefacts and associated Data Properties<br><br>Plans updated to account for quantity and complexity analysis |
| **Analyse Risks** | The required Data Safety Assurance Levels are established and justified<br><br>The required Data Safety Assurance Levels are analysed as part of system safety activities | Establish Data Safety Assurance Levels<br><br>Analyse DSALs as part of system safety activities | Data Safety Assurance Levels (DSALs)<br><br>Relationship between DSALs and other Integrity Levels | A DSAL (and supporting justification) for each identified risk |
| **Evaluate and Treat Risks** | Data Safety requirements are identified, documented and reviewed<br><br>Methods used to provide Data Safety assurance are defined and implemented<br><br>Evidence of compliance with the Data Safety requirements is documented, reviewed and approved | Review each risk and either: Avoid, Accept, Transfer, or Treat<br><br>Establish treatment methods for relevant risks<br><br>Implement and verify treatment methods | Methods and Approaches tables | A record of the agreed responses to each of the identified risks<br><br>Data Safety requirements that follow from these responses<br><br>A record of the treatment adopted for each of the identified risks<br><br>An assessment as to whether the risk has been suitably mitigated |

# 3  Establish Context

*It is a capital mistake to theorise before one has data.*
**Sherlock Holmes - "A Study in Scarlet" (Arthur Conan Doyle)**

## 3.1  Objectives

There are four objectives associated with establishing the context:

> 1. Key stakeholders and necessary approvers are identified.
> 2. Interfaces are defined and controlled.
> 3. The Data Safety assessment is planned.
> 4. Data Artefacts are identified.

## 3.2  Overview

This phase involves: understanding the context within which the system development occurs; understanding the system requirements; and understanding the system design. These factors help determine the risk appetite; that is, essentially, how much effort will be devoted to making risks as low as practicable. In turn, this will inform the nature and scope of assessments that are conducted during system development, introduction to service and operation. The factors also help identify stakeholders and approvers.

## 3.3  Activities

There are four activities associated with this phase:

### 3.3.1  Describe the organisational context

Part of this activity involves understanding the stakeholders involved, or with an interest, in the system. It is important to define how the stakeholders will interact and the derived requirements applicable to each stakeholder through interface control, similarly to systems engineering interface control procedures already in place in many industries. Consideration of external (e.g., economic, social, regulatory) and internal (e.g., culture, processes, strategy) factors is key to defining appropriate interface control measures. Note that, similarly to requirements, interface control may be iterative throughout the data safety assessment. In particular, interface control may need to be amended to take account of implemented data safety mitigations.

To form an understanding of each organisation's risk appetite, the Organisational Data Risk (ODR) assessment (Appendix B) can be used. The ODR assessment can be used at programme level to cover all stakeholders, or used by each individual stakeholder to determine the risk appetite appropriate to their managerial responsibility. Care should be taken when operating a modular approach to risk appetite to ensure that the resulting system can meet its cumulative requirements; that is, nothing has "fallen through the cracks". This is part of the requirements decomposition and verification / validation activities described later in the Data Safety Management Process.

Amongst other things, the ODR assessment includes: the severity of any potential accidents; organisational maturity; applicable legal and regulatory frameworks; and the size, complexity and novelty of the planned system. It results in a rating, from ODR0 (which corresponds to the lowest risk) to ODR4 (which corresponds to the highest risk). This rating provides an initial, top-level view of the magnitude of data-related risk. As such, it forms the basis for process tailoring; it also indicates the level of effort that should be allocated to data-safety issues.

To allow tailoring to be applied in cases where the ODR assessment has not been explicitly conducted, the following qualitative scale is also used:

| ODR Rating | Qualitative Description |
| --- | --- |
| ODR4 | High-risk |
| ODR3 | Medium-risk |
| ODR2 | |
| ODR1 | Low-risk |
| ODR0 | Very low-risk |

Note that, in the case of an ODR0, no further work is required.

In addition to the rating, the ODR can be used to prompt the identification of key stakeholders (i.e., those with an interest in the system) and necessary approvers (i.e., those who need to formally accept the system). Note that some approvers might be within the organisation, whilst others may represent external bodies. Likewise, approvers could also be customers or regulatory authorities.

Part of the process of establishing the internal context involves understanding organisational culture. A short Data Safety Culture questionnaire has been developed (Appendix C), which may help in this regard. This can be applied at an organisation level or, more likely, within an individual project team. The questionnaire could also be used to highlight the importance of data safety related issues within a project team. In addition, before and after measurements could be taken to establish the effectiveness of data safety related training.

## 3.3.2   Describe the system context

This activity is concerned with describing the system under analysis, as well as the key external influences on that system (examples of which include interfacing systems and human operators). There are obviously many aspects to this activity. For reasons of brevity only those aspects that are directly relevant to data safety are discussed here.

When describing the system it is often helpful to think in terms of producers and consumers of data. These may be external systems, or sub-systems, or a combination of both. In addition, it may be necessary to consider data supply chains, especially when there are a number of separate organisations involved. Note that the considerations discussed in this paragraph are intended to be addressed at a high level: the identification of specific pieces of data is a separate, but related, activity; the identification of required properties is another activity.

Also note that, as development progresses the system description is expected to be refined. This may enable the data safety system context to also be refined, supporting the data safety assessment planning.

### 3.3.3  Plan the assessment

This activity involves scheduling the phases associated with the Data Safety Management Process and acquiring the necessary resources to complete them. This also involves tailoring the generic process to meet the specific needs of a particular system development.

Details of the planned assessment may be recorded in a Data Safety Management Plan (DSMP). This could also be used to capture the scope of the analyses and the associated context. Together, this information constitutes the first section of the DSMP structure. Note that if a DSMP is created, it would be expected to be updated with details from subsequent phases. Alternatively, if a Safety Management Plan has already been developed for the project, data safety aspects may augment the existing Safety Management Plan.

Planning of the data safety assessment requires some knowledge of the quantity and complexity of data which requires assessment. Therefore, the DSMP (or Safety Management Plan) needs to be updated in subsequent phases once these details are known.

Planning of the assessment may be done through a procurement process. Procurers may wish to understand their potential supplier's understanding of data safety and plans to implement a data safety assessment. A Data Safety Supplier Questionnaire (Appendix D) has been developed to support this analysis; this questionnaire may also be used for auditing purposes.

### 3.3.4  Identify Data Artefacts

Data Artefacts are the key pieces of data that are generated, processed or consumed by the system. They provide the foundation for the remaining phases of data-related risk management.

To support the identification of Data Artefacts, a wide variety of Data Types has been enumerated. Cross-referencing these types against the system description should highlight the relevant artefacts.

Another way of identifying Data Artefacts involves considering the functions that the system performs and establishing the data that is required to support these.

A further option for confirming that all relevant Data Artefacts have been identified is to consider the different phases of the system lifecycle. This approach should help prevent an inappropriate focus on operational use of the system at the expense of, for example, artefacts associated with system test and evaluation.

## 3.4  Tailoring

The level of interface control required will depend heavily on the number of stakeholders, complexity of their interactions, and the contractual controls already in place. Many programmes already require Interface Control Documents (ICDs) to be developed for systems or equipment. The level of detail required in other programme interface control plans may be used as a guide for the requirements of data safety interface control.

The ODR assessment can be conducted at product line or individual product level, as appropriate for the organisation. It is generally not recommended to conduct the ODR without the context of a system type.

It is expected that the ODR assessment will be of most utility for organisations that do not have significant safety engineering experience and that are operating in less well-regulated industrial sectors. Conversely, an organisation that has considerable experience in the development of safety-critical systems in heavily regulated environments is unlikely to derive much benefit from the ODR assessment. Note that if this assessment is not conducted, some high-level qualitative estimate of risk may still be required (e.g., to support process tailoring); likewise, there will also be a need to identify key stakeholders and necessary approvers.

It is also expected that the Data Safety Culture questionnaire will be of most use for low-risk (i.e., ODR1) systems. In particular, it is expected that developers of higher risk systems will have extant processes to develop, maintain and monitor safety cultures.

The approach of including data-related aspects within a Safety Management Plan is recommended for complex or highly safety-critical systems. In this case the structure of the Safety Management Plan may be maintained, with the data safety assessment process being tailored to align to the overall safety assessment process.

The questionnaire that helps establish the level of "data maturity" in potential suppliers is expected to be of most use when new organisational relationships are being formed. Conversely, it may offer little value in situations where both organisations are familiar with each other, they have worked on data-related projects together before and there are suitable audit / review arrangements in place.

Data Artefacts may be defined at a number of levels. An artefact associated with a medical system could be described as "patient data". Alternatively, this could be split into smaller parts (e.g., "blood group"). Generally speaking, the highest possible level consistent with the system description should be used; this prevents an excessively long list of artefacts being developed. If necessary, those artefacts where further detail is needed can be refined as part of an iterative process that is focused on key issues.

Not every Data Type will be relevant to every system. Furthermore, for low-risk (ODR1) systems it may be sufficient to simply consider the groupings of types (e.g., "context", "implementation", etc.). Conversely, high-risk (ODR4) systems might need to consider every type, even if this results in a conclusion that a specific type is not relevant for the system in question.

A function-based approach to identifying Data Artefacts is likely to be enabled by design processes that also adopt a function-based perspective. If information from a function-based perspective is readily available then it should be used to support the identification of Data Artefacts. If this information is not readily available, it is recommended that it be generated for medium and high-risk systems (ie, ODR2 to ODR4, inclusive).

Considering data across the system lifecycle is a relatively simple activity, which is applicable to all systems (i.e., ODR1 to ODR4, inclusive).

## 3.5 Outputs

The key data-related outputs from this phase are:

1. A list of key stakeholders and necessary approvers for data-safety activities;

2. An interface control plan or list of control measures (possibly also a list of data owners, linked to the collection of Data Artefacts);

3. An estimate of the level of data-related risk (e.g., an ODR rating), along with supporting justification;

4. A plan for the remaining parts of the data safety assessment; and

5. A collection of Data Artefacts, described at an appropriate level of detail.

## 3.6    Guidance

### 3.6.1    Interface Control

The interfaces between data owners, and indeed data ownership itself, can be much more complicated that for hardware or software, where the owner can be clearly identified. Indeed, when combining items from various sources it is possible to create data for which there is not an "owner" in any traditional sense. In such circumstances it may be appropriate for the overall system owner to take responsibility for the collected data and, where appropriate, pass specific, formally-recorded requirements onto original data suppliers.

The data owners throughout the lifecycle of data within the system should be identified, or the lack of an owner highlighted where applicable, including where data is merged or modified through the system operation. This will facilitate a greater understanding of the "controllability" of data safety issues within the assessment at a particular organisational level.

### 3.6.2    Organisational Data Risk Assessment Form

The Organisational Data Risk (ODR) Assessment Form was generated to capture a high-level perspective on the risk posed to an organisation by data safety issues within a specific project. How it integrates with an organisation's existing risk (or safety) management processes is the responsibility of the implementing organisation. To facilitate this integration, the following paragraphs describe the connections between the ODR and the ISO 31000 standard for risk management. The ODR itself is presented in Appendix B.

Establishing the context of a risk assessment ensures that the system being considered and the scope of any assessment is well defined. This helps prevent an overrun of the assessment's boundaries and allows those items that are out of scope to be explicitly communicated to all stakeholders. In addition, it is the role of this activity to produce the risk criteria that a system will be judged on. The ODR assessment links directly to the sub-tasks identified by ISO 31000 for establishing the risk assessment context and introduces aspects to guide the assessor into focusing on data-specific risks.

Questions 2, 3 and 4 of the ODR align directly with establishing the external context of the risk assessment (Activity 5.3.2 from ISO 31000). They guide the assessor into judging the risk tolerance of external stakeholders, the level of risk that is allocated to the organisation and the regulatory environment within the project will operate.

Question 5 is concerned with establishing the internal context of the risk assessment (Activity 5.3.3), inviting the assessor to comment on the maturity of the organisation in terms of their attitude not just to risk, but specifically to data-driven risks.

Question 6 explores data ownership through the use cases of the system. This is related to the legal frameworks explored in Question 4, but also acts to lay the foundations of Activity 5.3.5, "Defining Risk Criteria", which requires an assessor to identify "the nature and types of causes and consequences that can occur and how they will be measured". This is expanded upon by Questions 1, 7 and 8 which go into data-driven specifics about failure consequences and the issues raised by data complexity, boundary complexity and system complexity for the project.

Finally, the scoring system of the ODR provides a heuristic for defining the risk criteria (Activity 5.3.5) which handles how to combine these different aspects of risk into a single, high-level estimate of the data-related risks associated with a given project. This means that the ODR can, for example, provide some guidance on the "4 + 1"st Data Safety Assurance Principle; that is, it provides some guidance on the amount of effort that should be directed towards explicit consideration of data safety issues.

It is of note that whilst the completion of an ODR fits within the context establishment activity it also augments the ongoing "communication and consultation" activity both by providing a standardised format for capturing the relevant information and securing endorsement.

### 3.6.3   Data Safety Culture Questionnaire

Part of the ODR assessment relates to assessing the organisation's maturity in managing data safety risks; responses are aimed at establishing the depth of awareness of data safety and the associated management processes within the organisation. However, measuring the level of awareness of processes and concepts in an organisation is not always easy. There may be sufficient high-level knowledge of this for the purposes of the ODR but it still may be an area that warrants further investigation.

To support this, a separate questionnaire has been developed to explore the specific area of measuring the data safety culture for a particular activity; whether this be for the organisation as a whole or for a particular project, service or activity. However, here the focus is on a personal view rather than a project or company view so the questionnaire would be completed by all, or a significant subset of, staff. Responses can be aggregated to give an overall data safety culture value. A key aspect of this approach is that it can be periodically repeated to determine trends: for example, if overall scores are declining, this may suggest that further training and briefings will be required.

More details on the data safety culture questionnaire are provided in Appendix C.

### 3.6.4   System Definition

The system under consideration should be understood and documented, including interfaces and safety-related data aspects. The process of documenting the system of interest furthers the understanding of stakeholders and approvers so they can make sensible judgements about the system. It also formally declares assumptions that are being made whilst assessing the system and clearly defines the limits of the assessment. In addition, different levels of risk may be associated with composites of safety-related data, which may be easier to manage than individual artefacts, or where independence cannot be demonstrated or maintained. Hence, the partitioning of datasets should also be considered during this phase.

### 3.6.5   Usage Scenarios

If safety-related data is incorrect it can become dangerous when used, either by making a computer or control system perform incorrect actions, or by misleading human users into making incorrect decisions. Since the danger can only be determined when the usage of the data is understood, risk assessment should involve both the consumer of the data and the producer.

The consumer assesses the use of the safety-related data. (In later phases of the Data Safety Management Process this information is used to define the required Data Properties: for example, how accurate a particular safety-related Data Artefact must be.)

The producer investigates how the safety-related data is collected and what errors might occur. (Building on activities in later phases of the Data Safety Management Process, the producer can provide some form of guarantee, or level of confidence, that the safety-related data meets the specific data-related requirements.)

In some cases a producer will be providing safety-related data without any knowledge of a specific user (e.g., mapping data or generic databases that are sold to many users). In these cases the producer will need to make some assumptions about possible users, and then clearly state what level of integrity the data has been produced to. It is then up to the users to check whether the declared integrity matches their need.

### 3.6.6  Supplier Data Maturity

As noted above, a number of usage scenarios involve data being supplied by subcontracted organisations. It is expected that some formal process will be used to select these suppliers. A questionnaire has been developed to help ensure that the supplier has suitable processes in place to manage data safety-related issues. This is available in Appendix D.

### 3.6.7  Data in System Lifecycles

Like other components of a safety-related system, the safety dependency of data is dictated by the context in which it is used and the causal links that become established where loss of one or more of the required properties can contribute to hazardous system states. For example, a given data set (say configuration data) could be used in a number of separate contexts such as:

- prototyping a system to demonstrate solution feasibility of a safety-related system;

- development testing of a safety-related system; and

- live operational use of a safety-related system.

In these cases, the data set is the same but the context of its use changes the safety significance and therefore the level of assurance that it may require. It follows that the assessed assurance level of a data set is also predicated on where and when in the lifecycle the data set will be applied.

To illustrate this concept, a number of generic model lifecycles are discussed below. Note that these are not intended to be prescriptive or mandate the use of any particular model. Instead, they are being used to illustrate how the Data Safety Management Plan could articulate these lifecycle considerations.

**Development** The following diagram represents a typical development lifecycle using an iterative development approach[1]. In this model there are key phases as the system transitions from concept through to testable executable code. The process is iterative in that several cycles of functional elaboration, design, development and test may be run and these typically will focus on the areas of the system that bear most technical risk or comprise the key functional use cases so the client gets early visibility of the system. This early awareness allows feedback to be provided into the next iteration to help steer the solution to the client's actual needs. Traditional waterfall implementation can map onto this model on the basis that there is only one iteration in each phase and all activities in one phase need to be completed before progressing to the next.

---

1    The diagram is based on IBM's Rational Unified Process (RUP), an iterative software development process framework. The original diagram is in the public domain.

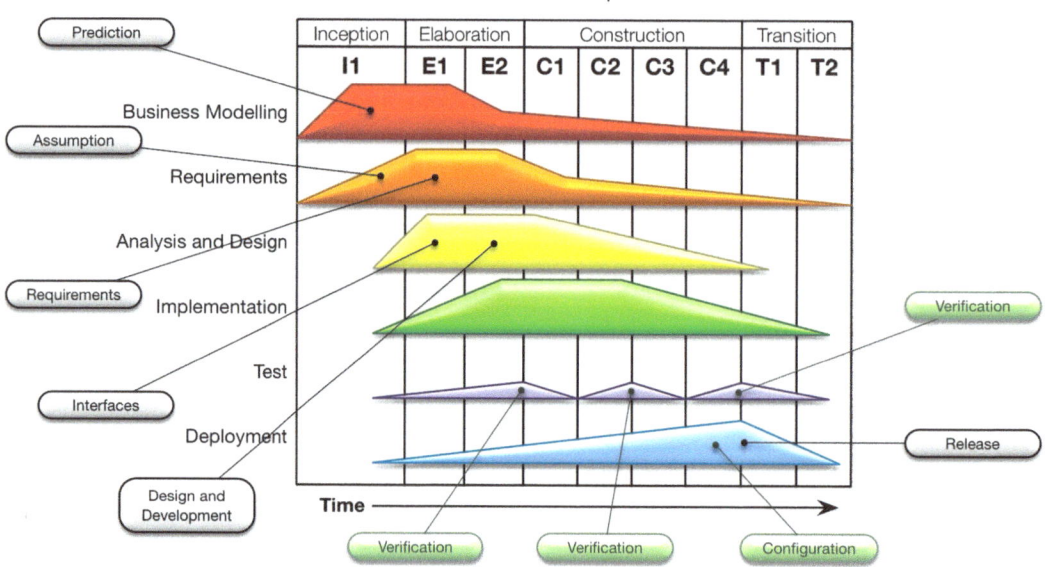

The model itself may vary depending on the specific needs of the project but the diagram illustrates that different Data Types become significant at different points of the process.

**Operational** Once a system has been developed it will move into an operational lifecycle or indeed, if data safety has not previously been considered for an enterprise, then the system could already be in operational use. These operational lifecycles tend to be cyclical in nature; the following diagram[2] illustrates a typical model.

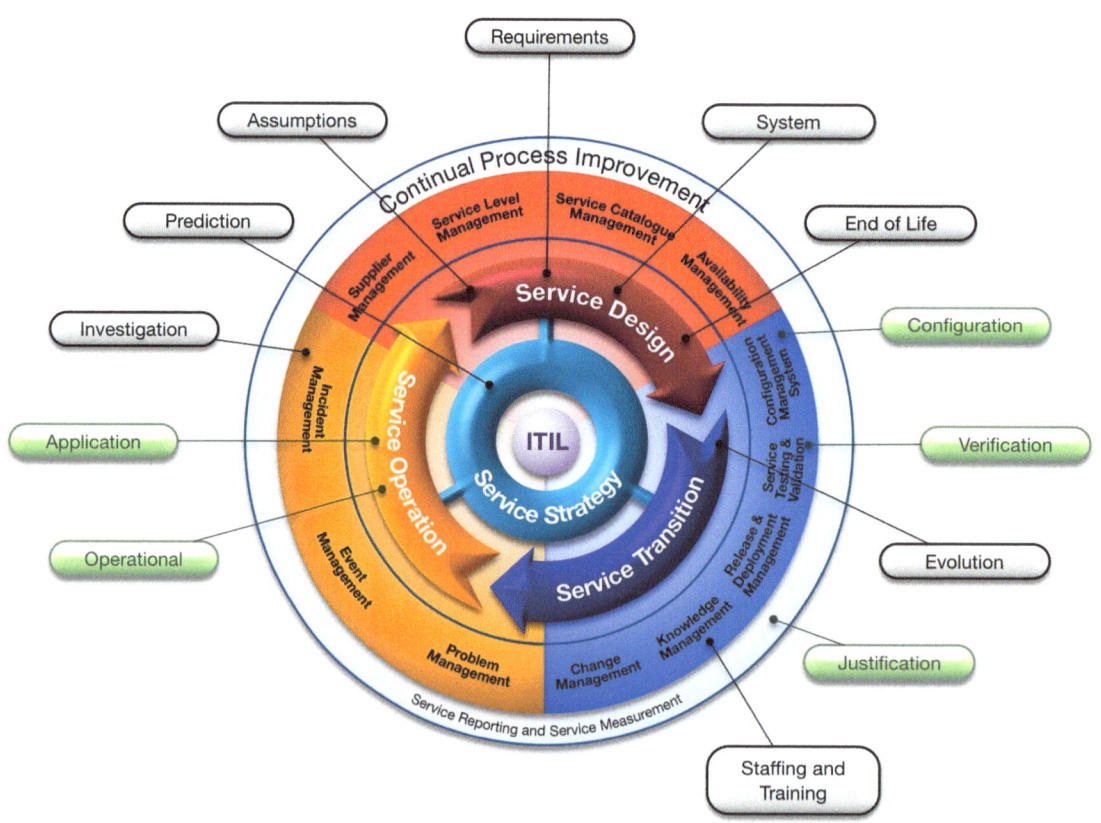

2    ITIL is a registered Trade Mark of AXELOS Limited. All rights reserved.

Again, specific Data Types will come into play at different periods in the process. Documenting the relationship between process steps and Data Types will therefore give clarity as to when a particular assurance technique needs to be applied.

**Data supply chains** The previous models relate to typical system supply and operate perspectives but there are also other data supply chains where a number of organisations engage in the procurement and use of safety-related data. These processes may include the development and operational lifecycles but a different model is required to fully represent the wider processes that are being employed. The following diagram shows such a model representing a data acquisition lifecycle.

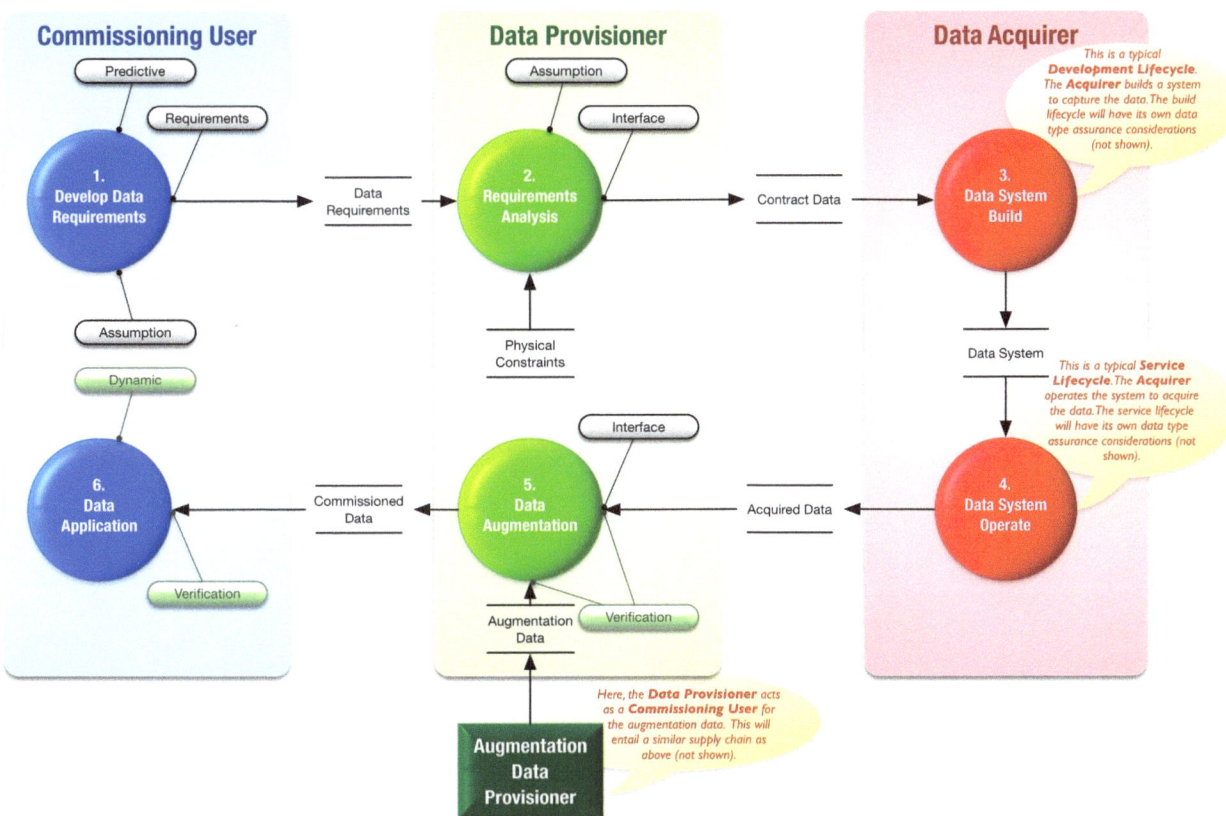

This model represents the interactions between three key organisations:

- The Commissioning User: the organisation that has the need for the data;

- The Data Provisioner: the organisation that will fulfil that need for data; and

- The Data Acquirer: the organisation employed by the Data Provisioner to carry out physical collection of data.

Note that these may be three separate organisations, or they may be separate business units within the same, larger, organisation.

In this supply chain, the Commissioning User is a Consumer of the data and the Data Acquirer is a Producer of data. The Data Provisioner acts both as a Consumer (from the Data Acquirer) and Producer (to the Commissioning User) of data. Similarly, an organisation that augments data sets is both a Consumer and Producer of data in the supply chain.

The Commissioning User Requirement Analysis is the key process step where the Commissioning User's expectations for data (including fidelity levels for associated properties) are agreed with the Data Provisioner. The requirements may be adjusted because of physical constraints (e.g., loss of precision because of physical measuring device constraints) and may include additional requirements to augment the captured data with additional information (e.g., airport codes added to a measurement of a given runway length).

The Data Provisioner may employ a Data Acquirer to capture the data (e.g., to carry out a physical survey of a site). The acquisition phase may itself require a specialised system to be built to perform the capture and data refinement to meet the Data Provisioner's specifications. Such systems will then themselves be subject to the Development lifecycle model considerations discussed above. Likewise, the data augmentation phase may require further system development processes or indeed, could trigger an instance of the model again as the Data Provisioner acting as a Commissioning User.

Acquired and augmented data is then fed into the operational system that has been built for providing the service of generating the commissioned data. This system in its service provision role would then typically follow the Operational Lifecycle process model discussed earlier.

### 3.6.8 Data Types

The full set of Data Types which can have safety implications is large: to date some twenty-three types (and one meta-type) have been identified.

The table below gives the current view of the types of safety-related data that contribute to, are used by, produced by or affected by safety-related systems. They are roughly organised into a number of categories, which aim to cover all aspects of the system lifecycle. Note that the list in the table below is non-exhaustive. Also note that a more detailed version of this table is available at Appendix E.

| No. | Type | Description |
|---|---|---|
| Context | | |
| 1 | Predictive | Data used to model or predict behaviours and performance |
| 2 | Scope, Assumption and Context | Data used to frame the development, operations or provide context |
| 3 | Requirements | Data used to specify what the system has to do |
| 4 | Interface | Data used to enable interfaces between this system and other systems: for operations, initialisation or export from the system |
| 5 | Reference or Lookup | Data used across multiple systems with generic usage |
| Implementation | | |
| 6 | Design and Development | Data produced during development and implementation |
| 7 | Software | Data that is compiled and executed to achieve the desired system behaviour |
| 8 | Verification | Data used to test and analyse the system |
| Configuration | | |
| 9 | Infrastructure | Data used to configure, tailor or instantiate the system itself |
| 10 | Behavioural | Data used to change the functionality of the system |
| 11 | Adaptation | Data used to configure to a particular site |
| Capability | | |
| 12 | Staffing and Training | Data related to staff training, competency, certification and permits |
| The Built System | | |
| 13 | Asset | Data about the installed or deployed system and its parts, including maintenance data |
| 14 | Performance | Data collected or produced about the system during trials, pre-operational phases and live operations |
| 15 | Release | Data used to ensure safe operations per release instance |
| 16 | Instructional | Data used to warn, train or instruct users about the system |
| 17 | Evolution | Data about changes after deployment |
| 18 | End of Life | Data about how to stop, remove, replace or dispose of the system |
| 19 | Stored | Data stored by the system during operations |
| 20 | Dynamic | Data manipulated and processed by the system during operations |
| Compliance and Liability | | |
| 21 | Standards and Regulatory | Data that governs the approaches, processes and procedures used to develop safety systems. |
| 22 | Justification | Data used to justify the safety position of the system |
| 23 | Investigation | Data used to support accident or incident investigations (i.e., potential evidence) |
| Meta-Property | | |
| +1 | Trustworthiness | (Meta) data which tells us how much the system can be trusted |

# 4   Identify Risks

> *I wanted to separate data from programs, because data and instructions are very different.*
> **Ken Thompson**

## 4.1   Objectives

There are two objectives associated with identifying the risks:

> 1. *Generic and historical examples of data-related issues are reviewed.*
> 2. *Risks are identified and linked to Data Artefacts and Data Properties.*

## 4.2   Overview

This phase involves identifying sources of risk and understanding the potential consequences; it should result in a comprehensive list of risks. From a system development perspective, these activities are likely to be concurrent with the development of more detailed system designs.

## 4.3   Activities

There are three, complementary, activities that can be used to identify risks. There is also an activity associated with updating planning documents.

### 4.3.1   Review the general, historical perspective

To help practitioners appreciate the different ways that data can lead to unsafe situations a generic list of typical ways that data can cause problems is available. This includes topics like ageing, biasing and the use of default values.

Further insight may be gained by reviewing historical accidents and incidents, a collection of which is included in this guidance document (Appendix A).

### 4.3.2   Conduct a top-down approach

If the system under consideration has clearly identified functions then data-related risks can be assessed by considering each function in turn and analysing what Data Artefacts and, more particularly, what Data Properties the function depends on.

If there are a limited number of safety-related functions, this is usually the simplest approach. This approach also has the advantage that it integrates well with other function-based, top-down approaches to assessing system safety.

### 4.3.3   Conduct a bottom-up approach

This approach starts from the Data Artefacts and explores the effects of data errors. In this context an error is a situation where a required Data Property is not exhibited. This may be achieved by a variety of methods, including a Hazard and Operability study (HAZOP).

### 4.3.4  Update planning documents

Once the data safety risks have been identified, the Data Safety Management Plan (or Safety Management Plan) requires updating to take account of the quantity and complexity of the analysis and mitigation activities needed to address the risks. While the DSMP may be updated throughout the data safety assessment, this is not necessarily essential for all projects.

## 4.4  Tailoring

The general, historical perspective review is a simple activity that does not require significant resources. Hence, it is recommended for all systems, regardless of risk level.

When conducting a bottom-up approach, it may not be appropriate to explicitly consider every possible property for every single artefact. In particular, for low-risk (ODR1) and medium-risk (ODR2 / ODR3) systems some form of tailoring may be expected: this may, for example, take the form of pre-selecting the properties that are most relevant, or limiting the layer of abstraction at which the system is considered (e.g., stop assessing at software module level, rather than reviewing lines of code).

Tailoring of the bottom-up approach may also be appropriate for some high-risk (ODR4) systems, but in this case an explicit argument that the tailoring has not adversely affected system safety would be expected. Furthermore, the risk identification process for high-risk (ODR4) systems is expected to be a highly structured affair. To support this, a number of data-related HAZOP Guidewords have been determined.

The top-down and bottom-up approaches provide different perspectives when attempting to identify data-related risks. For low-risk (ODR1) systems it may be appropriate to consider just one of these perspectives. Conversely, both perspectives would be expected to be considered (to some degree) for medium-risk (ODR2 / ODR3) and high-risk (ODR4) systems.

## 4.5  Outputs

The key data-related outputs from this phase are:

1. A description of the process used for risk identification;

2. A list of risks, linked to Data Artefacts and associated Data Properties; and

3. Plans that have been updated (to account for quantity and complexity of analysis).

## 4.6  Guidance

### 4.6.1  Historical Accidents and Incidents

Ideally, data safety risks would be identified and mitigated before they led to an accident or incident. However, this is not always the case. Historical occurrences can provide an indication of the data safety risks present in planned or existing systems. In particular, accidents and incidents can be analysed to identify potential contributory causes relating to data.

To support this type of analysis a number of previous accidents and incidents have been collected in Appendix A. These include, for example, cases that relate to the completeness, integrity and timeliness

Data Properties. They also highlight the importance of the adaptation Data Type and dangers associated with the inappropriate use of default data values.

Most of the current collection of accidents and incidents fall into three categories: aviation; maritime; and medical. However, the lessons that can be learned span a much wider range of application areas.

## 4.6.2   Ways that Data Can Cause Problems

There are some risk-inducing issues that are different or more prevalent for data than for other system elements. An incomplete collection of examples is provided below. This list may provide a quick way of identifying risks, which could be especially useful at an early stage of a project:

- **Fluidity** Hardware and software can undergo significant amounts of product assurance and once assured may change relatively infrequently. Where change is required to hardware or software, it can be carefully managed and the impact on the safety case appraised. This is not always the case for data, which is often much more fluid; indeed the ease with which data can be changed is one motivation for the move towards data-driven systems. This fluidity means that it is not always possible to revisit safety cases when data changes. Instead, the data fluidity, along with any associated safety impacts, may need to be captured in the system safety case. Fluidity can also provide a temptation for unscrupulous operators to falsify data, for example, after an incident has occurred. Rigorous configuration control procedures can help protect against this type of behaviour.

- **Reuse** For the purposes of this discussion, "reuse" is interpreted as use of the same data in a different system or system context (e.g., lifecycle phase). Just because data was valid for use in a particular system, it does not immediately follow that it can be reused again in a similar system. Many considerations associated with data reuse are similar to those of software reuse, for example: similarity of requirements; similarity of role in system; and similarity in required integrity / assurance level. One consideration that is different is that of timeliness: data that was valid for use in a particular system at a particular time is not necessarily valid for reuse in the same system at a different time.

- **Ageing** As highlighted above, all safety-related data has a lifetime and this needs to be explicitly managed. This can involve, for example, purging, deletion and alerting. It is also important to note that ageing can occur as a result of changes external to the system (e.g., the positions of other aircraft) or it can result from internal changes (e.g., in sensors monitoring system properties).

- **Transformation** Data is often filtered, mapped or aggregated as it moves through systems, sometimes creating new data sets as a result. Data Properties are not necessarily preserved by these processes. The main issues are loss of heritage / history / source information; data can also appear to become something else. Without careful management the integrity may become lowered to the lowest common denominator and this needs to be recognised. Additional checks (e.g., validation checks, sanity checks) or assurance measures may need to be put in place to ensure that required integrity / assurance is maintained.

- **Ownership** The transformation of data can result in a lack of clarity regarding who has ownership of, and responsibility for, the data (if anyone). It is important that responsibility for errors can be tracked, for example, to determine whether they were present in the initial data or whether they arose as part of the transformation process. Establishing clear roles within the data supply chain can help mitigate these issues.

- **Archiving and Retrieval** Safety-related data needs to be available when required. There is thus a need to think about data accessibility over the complete system lifetime. It is also important to consider what properties of the data need to be preserved and how this affects the choice of storage medium.

- **Biasing** This is a systemic inaccuracy in data due to the characteristics of the process employed in the creation, collection, manipulation, presentation and interpretation of data. It is usually an unintentional distortion in the data set. Although there is no perfect way of checking for this within the system, completeness, statistical and validity checks on data sets may help.

- **Defaulting** Many systems use default or initial values for data items; sometimes in data sets and sometimes embedded in software. Often these default values are designed to be neutral (e.g., "0") or unrealistic (e.g., "VOID"). There are essentially two cases: (i) initialisation data which may persist and be mistakenly taken as a real value when in fact it should have been changed; (ii) data that is used when no meaningful value has been assigned (e.g., during data migration or data exchange between systems). These issues can often be managed through good design of data structures, for example by the inclusion of a validity flag.

- **Sentinels** A sentinel value is a data value that is used to indicate a special action needs to be taken, typically indicating the end of a record or a data set. The sentinel value should be one that is not allowable in the data set itself, but often is not properly considered and may use common sequences (e.g., five zeroes). Sentinels can cause problems in two ways: (i) where they are not recognised and so, for example, processing continues past the sentinel; (ii) where the data itself somehow contains the sentinel value and so processing is erroneously interrupted. Sentinels can be a particular risk in long-lived systems and data sets. As with the issue above, the management or elimination of this issue may often be achieved through improved data structures.

- **Aliasing** This is an effect that causes different data to become indistinguishable when accessed; that is, there is only one record when there should be several. This could be due to the way the data is filtered, sampled, indexed, stored or retrieved. The data issues are typically related to loss of resolution leading to similar data points appearing to be identical. Hence, methods to maintain resolution, including use of unique indexes, may be beneficial.

- **Disassociation** This effect is, in some senses, the opposite of aliasing; there are several records when there should only be one. This could occur, for example, if two records are created for the same individual using slightly different names. It could also arise if different systems use different indexing methods and the association between the indexes becomes corrupted. Again, methods to maintain data resolution can be beneficial.

- **Masking** This issue can arise if a notable proportion of a data set is of a poor quality. This poor quality data can mask errors in the way that the system handles the good quality data. One way of protecting against this issue is the generation and use of test sets of appropriate size and quality, although form some applications this may be a non-trivial task.

### 4.6.3  Data Properties

Data Properties are used to establish what aspects of the data (e.g., timeliness, accuracy) need to be guaranteed in order that the system operates in a safe manner.

James Inge's work [18] produced a useful taxonomy of data types, and went on to look at faults in data. He concluded that a rigid taxonomy of data types was unhelpful due to various properties, or characteristics, of the data which vary independently. In short, it is the combination of Data Type with the required Data Properties that facilitates safety analysis.

Data Types were discussed in the preceding phase. A collection of Data Properties has been produced; this is documented in the following table. Typically speaking, it is the loss of one of these properties that presents a hazard. Note that, this notion of "loss" is dependent on the intended use: for example, what is "timely" for one use may not be for another. Also note that this list is non-exhaustive.

| Property | Description |
|---|---|
| Integrity | The data is correct, true and unaltered |
| Completeness | The data has nothing missing or lost |
| Consistency | The data adheres to a common world view (e.g., units) |
| Continuity | The data is continuous and regular without gaps or breaks |
| Format | The data is represented in a way which is readable by those that need to use it |
| Accuracy | The data has sufficient detail for its intended use |
| Resolution | The smallest difference between two adjacent values that can be represented in a data storage, display or transfer system |
| Traceability | The data can be linked back to its source or derivation |
| Timeliness | The data is as up to date as required |
| Verifiability | The data can be checked and its properties demonstrated to be correct |
| Availability | The data is accessible and usable when an authorized entity demands access |
| Fidelity / Representation | How well the data maps to the real world entity it is trying to model |
| Priority | The data is presented / transmitted / made available in the order required |
| Sequencing | The data is preserved in the order required |
| Intended Destination/Usage | The data is only sent to those that should have them |
| Accessibility | The data is visible only to those that should see them |
| Suppression | The data is intended never to be used again |
| History | The data has an audit trail of changes |
| Lifetime | When does the safety-related data expire |
| Disposability / Deletability | The data can be permanently removed when required |

## 4.6.4 HAZOP Guidewords

A Hazard and Operability study (HAZOP) provides a structured approach for identifying hazards. It involves a multidisciplinary team collaborating to identify potential hazards and operability problems. Structure and completeness are supported through the use of guideword prompts, for example, considering the implications if software components perform functions early, late or not at all. These prompts are intended to stimulate imaginative thinking, to focus the study and to elicit ideas and discussion.

The following table lists a set of guidewords for a data-focused HAZOP. The list is non-exhaustive. Other guidewords may be useful for particular systems, or may be used to ensure the data safety assessment is fully integrated within the system safety assessment. Note that a more detailed version of this table, including specific HAZOP Data Guidewords, is available at Appendix F.

| Property | HAZOP Data Properties |
|---|---|
| Integrity | Loss, partial loss, incorrect, multiple |
| Completeness | Loss, partial loss, incorrect, multiple |
| Consistency | Loss, partial loss, incorrect, multiple, too early, too late, loss of sequence |
| Continuity | Loss, partial loss, incorrect, late, loss of sequence |
| Format | Loss, partial loss, incorrect, multiple |
| Accuracy | Loss, partial loss, incorrect, multiple |
| Resolution | Loss, partial loss, incorrect, multiple |
| Traceability | Loss, partial loss, incorrect, multiple, too early, too late, loss of sequence |
| Timeliness | Loss, partial loss |
| Verifiability | Loss, incorrect, partial loss, multiple, too early, too late, loss of sequence |
| Availability | Loss, partial loss, multiple, too early, too late |
| Fidelity / Representation | Loss, incorrect, partial loss, multiple, too early, too late |
| Priority | Loss, incorrect, partial loss, multiple, too early, too late |
| Sequencing | Loss, incorrect, partial loss, multiple |
| Intended Destination / Usage | Loss, incorrect, partial loss, multiple, too early, too late, loss of sequence |
| Accessibility | Loss, incorrect, partial loss, multiple, too early, too late |
| Suppression | Loss, incorrect, partial, too early, too late, too much, too little |
| History | Loss, incorrect, partial loss, multiple |
| Lifetime | Loss, too early, too late, incorrect, multiple, loss of sequence |
| Disposability / Deletability | Loss, incorrect, partial, too early, too late |

# 5    Analyse Risks

> *I love data. I think it's very important to get it right, and I think it's good to question it.*
> ***Mary Meeker***

## 5.1    Objectives

There are two objectives associated with analysing the risks:

> 1. The required Data Safety Assurance Levels are established and justified.
> 2. The required Data Safety Assurance Levels are analysed as part of system safety activities.

## 5.2    Overview

This part of the risk management process involves developing an understanding of the consequences and likelihood of each risk. From the perspective of safety-critical and safety-related systems this understanding allows System (or Safety) Integrity Levels or Development Assurance Levels to be determined. Likewise, this understanding should be used to allocate Data Safety Assurance Levels.

## 5.3    Activities

There are two activities associated with this phase.

### 5.3.1    Establish Data Safety Assurance Levels

The key activity in this phase is to establish the (untreated) likelihood and severity of each risk identified in the preceding phase.

In order to analyse risks and, more particularly, to align data safety with other risk management processes, there is a need to overcome problems stemming from the use of term "likelihood" in a situations where there may be no failure rates. For this reason the Data Safety Assurance Level (DSAL) was developed. The DSAL metric is not a statistical measure of likelihood, or a literal numeric measure of integrity. Instead, the DSAL metric is an indicator for the level of rigour that an assurance argument requires. As such, DSALs share a common theoretical basis with concepts like Item Development Assurance Levels [31] and development process systematic capability [26].

DSALs are measured on a scale of DSAL0 (lowest-assurance) to DSAL4 (highest-assurance). They are allocated as indicated in the following table.

| Severity | Likelihood | | |
|---|---|---|---|
| | **High** | **Medium** | **Low** |
| Negligible | DSAL0 | DSAL0 | DSAL0 |
| Minor | DSAL1 | DSAL0 | DSAL0 |
| Moderate | DSAL2 | DSAL1 | DSAL1 |
| Major | DSAL3 | DSAL3 | DSAL2 |
| Catastrophic | DSAL4 | DSAL4 | DSAL3 |

It is possible that the additional understanding developed during this part of the process may mean some previously identified Data Artefacts are no longer of consequence; similarly, it is possible that this process may identify additional artefacts or a need to refine the description of existing artefacts.

### 5.3.2   Analyse DSALs as part of system safety activities

Allocating a DSAL is a significant part of controlling data safety risks, but it is not the only part. It is important that DSALs are considered as part of wider system safety activities, rather than being viewed as a separate item.

For medium-risk (ODR2 / ODR3) and high-risk (ODR4) systems it is likely that integrity, or assurance, levels will be calculated from perspectives other than data safety. Possible examples include Safety Integrity Levels from IEC 61508 and Item / Function Development Assurance Levels from Aerospace Recommended Practice (ARP) 4574A.

This activity involves comparing DSALs with these other integrity, or assurance, levels. Assuming a typical scenario of a system processing or manipulating data flowing through it, there are two cases to consider:

1.  Can the data affect the software? In particular, this question is concerned with whether the data can affect the software such that the safe operation of the system is jeopardised. Obviously an ideal system would be able to handle any data fed into it safely without problems, but this is often not the case. An example might be a legacy system which has limited error checking and so may fail in unsafe ways if fed with data which is outside of the expected range. Formally this question can be stated as: *Given a system containing software written to a particular Software Assurance Level (which may be none), what should the DSAL of the processed data be to preserve correct operation of the system?*

2.  Can the software affect the data? In particular, this question is concerned with whether the system's software can affect the data being processed or manipulated in such a way that Data Properties that are important for safety might be lost. Some examples might be systems which transform messages, losing any associated checksum protections, thereby possibly affecting the integrity of the data within the message; as a minimum, this removes a means of checking data integrity. Another example might be a system that can delay data flowing through it, (e.g., due to buffering) when timely delivery of the data is critical. Formally, this question can be stated as: *Given data at a particular DSAL, what should the Software Assurance Level of the software in the system be in order to preserve the DSAL of the data?*

Guidance is provided for both of these cases.

## 5.4   Tailoring

To facilitate integration with other safety assessment processes, the severity and likelihood definitions (stated in the following guidance) may be replaced with other definitions already in use within an organisation.

DSALs can be applied at different levels and to different constructs. For example, in the case of simple, low-risk systems it may be appropriate to apply a single DSAL to an entire system. Alternatively, it may be appropriate to apply DSALs to sub-systems, for example, to match the level at which other integrity, or assurance, levels have been determined.

Another option is to apply DSALs to Data Artefacts rather than directly to risks. This approach has the advantage that treatments are often related to artefacts; it can work well where there is a simple relationship between artefacts and risks.

## 5.5    Outputs

The key data-related output from this phase is:

1.    A DSAL (and supporting justification) for each risk identified in the previous phase.

## 5.6    Guidance

### 5.6.1    Establishing DSALs

The likelihood of a data-safety related risk is qualitatively determined by considering the following characteristics:

1.    Proximity: how directly a data failure will lead to an accident;

2.    Dependency: how dependent the application is on the dataset;

3.    Prevention: the ability of the systems architect / developers to guard against errors;

4.    Detection: the likelihood of being able to detect a data failure prior to an accident; and

5.    Correction: the ability of the system to work around or correct errors.

For example, errors that are easy to guard against are associated with low likelihoods. Conversely, errors that are difficult to detect are associated with high likelihoods. The following table illustrates how aspects of these characteristics map to three, qualitative likelihoods.

| | Likelihood | | |
|---|---|---|---|
| | **High** | **Medium** | **Low** |
| Proximity | A known use of the data is highly likely to lead to an accident. | A possible use of the data could lead to an accident. | All currently foreseen uses of the data could lead to harm only via lengthy and indirect routes. |
| Dependency | Data is completely relied upon. | Data is indirectly relied upon. | Little reliance on data. |
| Prevention | Difficult or impossible to guard / barrier against errors. | Possible to guard / barrier against errors. | Easy to guard / barrier against error. |
| Detection | Low or no chance of anything else detecting an error. | Some other people / systems are involved in checking the data. | Many other people / systems are involved in checking the data. |
| Correction | Difficult or impossible to correct or workaround errors. | Possible to correct or workaround errors. | Easy to correct or workaround errors. |

When applying this table to a specific data-related risk it is likely that consideration of different characteristics will result in different likelihoods. In order to provide an overall likelihood, it is assumed that the actions implied in the above table are taken. For example, a low likelihood for the "prevention" characteristic is only valid if the easy guard / barrier is actually implemented. Similarly, it is assumed that if an error is found, under the "detection" characteristic, an appropriate response is implemented. With those assumptions in place, the overall DSAL is the lowest likelihood of any characteristic.

Risk severity is estimated against a five-point scale, as indicated in the following table.

| Severity | Description |
| --- | --- |
| Negligible | Negligible harm. Negligible environmental impact. |
| Minor | Minor injury or temporary discomfort for 1 or 2 people. Minor environmental impact. |
| Moderate | An accident resulting in minor injuries affecting several people or one serious injury. Some environmental impact. |
| Major | A serious accident resulting in serious injuries affecting a number of people, or a single death. Major environmental impact. |
| Catastrophic | An accident resulting in several deaths. The accident could affect the general public or have wide and catastrophic environmental impact. |

As noted earlier, DSALs have some commonality with things like (Item / Function) Development Assurance Levels (IDALs / FDALs). However, this commonality does not extend across all aspects. For example, there is an accepted calculus of FDALs in which two independent lower-integrity functions can be used to replace a single higher-integrity function. There are two reasons why this type of calculus is not appropriate for DSALs:

1. The definition of a DSAL already caters for interactions. For example, using two independent Data Artefacts to provide similar information to a system function reduces the "dependency" of each artefact.

2. These types of consideration are most closely related to system architecture, from which Data Artefacts, associated Data Properties and risks are derived. Hence, rather than applying any calculus at the DSAL tier it is more appropriate to apply this to, for example, FDALs, with DSALs changing as a consequence of the revised system definition.

### 5.6.2  Analysing DSALs

In both cases (i.e., data affecting software and software affecting data) the degree of contribution and existing mitigation position are important. The mitigations should be proportionate and full credit for existing mitigations may reduce or obviate the need for additional work. A particular case is the use of strong checksums to "wrap" the data. If these are preserved through processing and can be checked later on, then undetected corruption situations can be largely discounted.

With regards to the first case (i.e., **data affecting software**) the issue is that the data used by the system may affect the software execution in a way that could credibly lead to hazards, but only where this data-induced effect is not easily detected or mitigated by other means. If this is the situation then appropriate measures need to be put in place *within the data* to mitigate this risk. Alternatively (or additionally), if the software within the system can be modified, mitigations could be placed *within the software* to achieve or enhance the mitigation needed:

1. **Mitigation within the data** In many cases the best or only option is to improve the quality of the data to avoid the issue. This can be done by introducing a DSAL for the data, related to the severity of the hazard which may be induced, and thereby addressing the issue at cause. In this

case, the DSAL is bearing a large amount of responsibility. This may mean that a greater number of the "Recommended" risk treatment methods and approaches (identified in the following phase) need to be implemented.

2. **Mitigation within the software** If the data cannot be assured to a DSAL (e.g., if it is supplied by a third party or legacy system), sometimes changes can be made to the software in the system to improve the situation. These mitigations could be functional (e.g., introduction of better range checking, rejection of illegal combinations of data values). Of course this requires not only specific software changes but also associated verification of these new features and any software changes will have to be implemented to an appropriate Software Assurance Level. However there may be no particular functional mitigation or set of mitigations that can be targeted at the particular issue (e.g., testing for illegal combinations of values is too complex). In this latter case, potentially the complete set of software within the system responsible for manipulating the data should be developed (or re-developed) to a suitable Software Assurance Level. This level should be related to the severity of hazards it is mitigating.

With regards to the second case (i.e., **software affecting data**) the issue is that the software may affect (e.g., corrupt, delete) the data in a way that key safety properties may be lost and, furthermore, that loss may not be easily detected. In general the key Data Properties should be considered to see if any important ones for this data may be jeopardised. If so, a Software Assurance Level from an appropriate standard or guideline should be introduced that mitigates this risk of undetected property loss. This Software Assurance Level should be determined by the hazards that could be caused, and may be localised to the software that can cause the problem.

However there are specific functional and architectural approaches that may reduce or avoid the need for a Software Assurance Level, including use of strong checksums and digital signatures, as well as techniques such as storing multiple copies, independent channels and so on. However it is important that the software performing the check of the Data Property will itself need to be developed to the introduced Software Assurance Level. The key is to establish what the software in the system is doing to the data: if the operations are simple and non-changing, then the risk is lower; if the operations are complex involving transforming the data, using the data to calculate and insert new values, or reformatting the data then the risk is higher.

It is necessary to decide how effective the functional mitigations are in reducing impact to the particular data properties, e.g., a strong checksum may be very effective at detecting unwanted change and therefore lower the Software Assurance Level (from the perspective of data-related requirements). However a checksum may not help at all if the issue is one of timely message delivery. More information on the use of checksums in the aviation domain is available in [14]; this information is likely to be applicable to other domains as well.

# 6 Evaluate and Treat Risks

> *In any collection of data, the figure most obviously correct, beyond all need of checking, is the mistake.*
> ***Finagle's Third Law***

## 6.1 Objectives

There are three objectives associated with evaluating and treating the risks:

> 1. Data Safety requirements are identified, documented and reviewed.
> 2. Methods used to provide Data Safety assurance are defined and implemented.
> 3. Evidence of compliance with the Data Safety requirements is documented, reviewed and approved.

## 6.2 Overview

This phase involves deciding, at a generic level, what action (if any) should be taken for each of the risks identified in preceding phases. This decision will be influenced by the organisation's risk appetite and other factors determined as part of the Establish Context phase. From some perspectives it may seem strange that, as indicated in the first objective, requirements are identified at such a late phase. This is a consequence of explicitly linking data safety requirements to risks associated with Data Properties of Data Artefacts, and the use of Data Safety Assurance Levels to describe levels of rigour.

This phase also involves identifying, implementing and verifying treatments for the risks emerging from the previous phase. Part of verifying the treatment involves checking technical details of the chosen approach; another part involves re-assessing the post-treatment risk to determine whether it is now acceptable.

## 6.3 Activities

This phase involves reviewing each risk, including the associated DSAL, and determining the appropriate response. Essentially, this phase aims to answer the question: can we accept this risk or does some action need to be taken? This is likely to require discussion amongst a number of stakeholders. From a system-safety perspective, there is nothing intrinsically special about data-related risks. Hence, it is recommended that evaluation of data-related risks be conducted alongside the evaluation of other system risks, as part of an organisation's standard risk evaluation process.

Typically responses will be one of:

- **Avoid** In this case the risk is considered so significant that it cannot practically be mitigated. In such circumstances it may be appropriate to avoid the risk (e.g., by stopping the project, or embarking on a significant redesign).

- **Accept** In this case there is no need for further action. This may be suitable for DSAL0 risks; if it were used for other DSALs then appropriate justification is likely to be required.

- **Transfer** In this case the risk is transferred to another organisation. This can be achieved, for example, by placing specific requirements on external suppliers of data (i.e., data producers). It is important that any such risk transfers are formally documented, understood and accepted by both parties.

- **Treat** In this case there is a desire to reduce the risk. This can be achieved by reducing either the severity or the likelihood. Choosing this response often involves having an outline view of how the risk may be reduced.

In addition to deciding on and documenting the appropriate response to each risk, this phase also includes gaining approval for these decisions.

In cases when a decision is made to treat a risk, suitable methods and approaches should be identified. To assist in this process, a range of potential methods and approaches are included in this guidance. These are mapped against DSALs, Data Properties and a selection of lifecycle data types.

Once a treatment method has been established and implemented there is, of course, a need to determine whether the expected mitigation has been achieved. Equivalently, there is a need to consider whether the residual risk may now be accepted (or whether another one of the responses identified above is necessary).

## 6.4   Tailoring

It is expected that records will be kept of the discussions that occur as part of risk evaluation. For low-risk (ODR1) systems, this may be in the form of a brief memo. Conversely, for high-risk (ODR4) systems, a detailed, structured record, which is placed under formal control, may be expected; in this case these discussions may be recorded as part of the system's Hazard Log or as part of a Data Safety Management Plan.

DSALs can be applied at varying levels of abstraction. For small-scale or low-risk (ODR1) systems it may be appropriate to consider treatments at higher levels of system abstraction. For example, this could involve applying a single DSAL to a sub-system or even to the system in its entirety. The latter approach could be appropriate where the data in the system interacts in complex ways and the associated safety risk does not warrant a detailed investigation of these interactions.

A significant amount of tailoring is implicit in the way the tables of methods and approaches are constructed. At best a method / approach may be Highly Recommended as a way of maintaining a required Data Property at a given DSAL. In addition, the tables are not exhaustive; additional, or alternative, methods and approaches can be used.

## 6.5   Outputs

The key data-related outputs from this phase are:

1. A record of the agreed responses to each of the identified risks, along with any supporting justification;

2. Data safety requirements that follow from these responses;

3. A record of the treatment adopted for each of the identified risks, including evidence that the treatment has been successfully implemented; and

4. An assessment as to whether the risk has been suitably mitigated (and, if not, plans for further mitigation activities).

## 6.6   Guidance

### 6.6.1   Risk Treatment

There are a range of approaches that could be used. One option might be to redesign part of a system, either to remove the risk or to incorporate safety devices. Alternatively, ways of mitigating (or, more generally, treating) the risk could be devised. Another option could be to conduct further analysis, for example to better understand the likelihood of a risk occurring. In extreme cases, risk evaluation may lead to a recommendation to cancel a project.

It is apparent that some of these approaches involve repeating activities (or part of activities) discussed in earlier sections of this document. This type of repetition is to be expected given the iterative nature of risk management.

Discussion is an important part of risk evaluation, allowing a variety of different perspectives to be brought to bear. Documentation is also important, partly to allow these discussions to occur on an even footing and partly to ensure that decisions and supporting rationale are recorded.

### 6.6.2   Mitigating Data Safety Risks

A range of methods and approaches can be used to mitigate the identified data safety risks. Since mitigation can be a complex process, requiring collaboration with all system engineering elements, a collection of high-level mitigation measures is provided; these may particularly assist those attempting to explain the process to non-practitioners, or those conducting assessments in less regulated environments. For practitioners assessing systems which do not have a high safety criticality, these high-level mitigation measures may prove sufficient.

For practitioners conducting assessments in highly regulated environments, or for highly safety-critical systems, sets of appropriate mitigation measures should be derived from the high-level table. To assist these practitioners, suggested methods and approaches are provided in a series of more detailed tables. These methods and approaches have been developed through cross-industry collaboration, but they may not be complete, especially for different types of system.

The practitioner should always consider whether the mitigation measures used to mitigate the data safety risks are sufficient for their purposes. In highly safety-critical systems, each data safety risk can be linked to a system-level hazard (or a new system-level hazard identified). Each hazard is tracked in accordance with the existing system safety methodology and mitigations are reviewed in terms of their feasibility, potential to introduce new or amended hazards, and effectiveness.

#### 6.6.2.1   High Level Mitigation Measures

The following tables provide high level mitigation measures for data safety issues. Each of these mitigation measures should be reviewed to establish whether it is relevant to the system under assessment.

For each Data Safety Assurance Level, the tables indicate whether the method / approach is:

- Highly Recommended (HR);

- Recommended (R); or

- No recommendation for or against being used (-).

| Ref | Mitigation Measure | DSAL | | | | Comment |
|---|---|---|---|---|---|---|
| | | 1 | 2 | 3 | 4 | |
| M_01 | Documentation of data context and suitability for use | HR | HR | HR | HR | e.g., data flow diagram to document and agree how data is handled in the system, recording the impact of design decisions on the data aspects of the safety case |
| M_02 | Definition of data ownership through the data lifecycle in the system | R | R | HR | HR | e.g., governance model, interface control document |
| M_03 | Definition and traceability of data requirements | HR | HR | HR | HR | e.g., requirements management, use of test data / test cases |
| M_04 | Recorded trustability of the data source(s) | R | R | HR | HR | e.g., source of the data is trusted (with 'trusted' to be defined in detail for the system), or there are multiple sources of data which are correlated |
| M_05 | Editing limitations | R | R | HR | HR | e.g., encapsulation of data, access limitations |
| M_06 | Diverse and / or redundant manipulation of data | R | R | HR | HR | e.g. data partitioning separation of data that is managed differently (architectural decisions) |
| M_07 | Automatic system checking functionality | - | R | HR | HR | e.g., Built-In Test (BIT), heartbeat functionality |
| M_08 | Monitored, controlled, or redundant manipulation of data | - | - | R | HR | e.g., redundant channels processing the data as hot standby |
| M_09 | Diverse and / or redundant storage of data | R | R | HR | HR | e.g., redundant storage of data, multiple different media types used to back-up the data |
| M_10 | Data recovery mechanisms | R | R | HR | HR | e.g., backward recovery, error correcting codes |
| M_11 | Tracking of data | R | R | HR | HR | e.g., digital signatures, sequence numbers, logging of data processing events, using metadata, configuration management |
| M_12 | Recorded derivation of test data | R | R | HR | HR | e.g., test data derived from an established system and supported by field evidence, or from another 'trusted' source |
| M_13 | Documented compliance against the data requirements | HR | HR | HR | HR | e.g., use of test data / test cases |

### 6.6.2.2   Detailed Methods and Approaches

The following collection of tables detail methods and approaches which may be used by practitioners conducting safety assessments on highly safety-critical systems. The tables map the methods and approaches to Data Types. To aid legibility, these data types are abbreviated using the scheme shown in the following table.

| Data Type | Abbreviation |
|---|---|
| Verification | V |
| Infrastructure | I |
| Dynamic | D |
| Performance | P |
| Justification | J |

The tables also map the methods and approaches to the Data Properties. To aid legibility, these properties are abbreviated using the scheme shown in the following table.

| Data Property | Abbreviation |
|---|---|
| Integrity | I |
| Completeness | C |
| Consistency | N |
| Continuity | Y |
| Format | O |
| Accuracy | A |
| Resolution | R |
| Traceability | T |
| Timeliness | M |
| Verifiability | V |
| Availability | L |
| Fidelity / Representation | F |
| Priority | P |
| Sequencing | Q |
| Intended Destination / Usage | U |
| Accessibility | B |
| Suppression | S |
| History | H |
| Lifetime | E |
| Disposability / Deletability | D |

In an attempt to aid usability, these detailed methods and approaches tables have been organised into eight, loosely-defined categories:

- System Design;

- Data Design;

- Data Implementation;

- Data Migration;

- Data Testing;

- Test Data;

- Media - Paper; and

- Media - Electronic.

### 6.6.2.3 System Design

| Technique | Data Types | DSAL | | | | Notes | Data Property |
|---|---|---|---|---|---|---|---|
| | | 1 | 2 | 3 | 4 | | |
| Built-in-Test / Built-in-Test Equipment (BIT / BITE) | ..D.. | - | R | HR | HR | Application tests the data (e.g., at start-up or when requested by an operator). | IC.......V.......... |
| Cyclic / Continuous BIT | ..D.. | - | - | R | HR | Application applies tests to the data it is processing continuously (e.g., for a live data stream) or periodically (e.g., every nth message, every hour). | IC.Y.....VL......... |
| Backward recovery | ..D.. | R | R | HR | HR | If a fault in data has been detected, the system resets to an earlier internal data set, which has been proven consistent. | IC................. |
| Parity Checks | ..D.. | R | R | HR | HR | Within data, e.g., Hamming codes, Reed-Solomon, Hagelbarger. | I.................. |
| Automatic Error Correction | ..D.. | R | R | HR | HR | Detected errors are corrected automatically. | IC................. |
| Checksums / Cyclic Redundancy Checks (CRCs) / Hashes | ..D.. | - | R | HR | HR | Digests of datasets are produced, included with the dataset and checked to provide confidence that the data is unaltered. | IC................. |
| Digital Signatures | ..D.. | - | R | HR | HR | For non-repudiation and integrity of data. | I......T......U..... |
| Sequence Numbers | ..D.. | R | R | HR | HR | Data bears sequence numbers so the integrity of a data stream can be checked (e.g., monotonic increase, duplicate detection). | ICN.........PQ...... |
| Automatic Repeat Request | ..D.. | R | R | HR | HR | Automatic Repeat-reQuest (ARQ) to repeat transmission of data which has not been received correctly. | IC.Y............... |

| Technique | Data Types | DSAL | | | | Notes | Data Property |
|-----------|------------|------|---|---|---|-------|---------------|
| | | 1 | 2 | 3 | 4 | | |
| Auditing Facilities | ..DP. | - | R | HR | HR | Changes to data properties are audited so the before and after values are recorded and also other related information such as the author and the time of the change. | .......T.V.......H.. |
| Logging Facilities | ..DP. | R | R | HR | HR | Data processing events are logged to allow support staff to monitor the health of the system and provide diagnostic information. | .......T.........H.. |
| Encapsulation | ..D.. | R | R | HR | HR | The hiding of data so that it is only accessible through well defined interfaces. | .............UB.... |
| Multiple Stores | ..D.. | - | - | R | HR | The same instance of a data set or data items is stored in multiple locations. | ...............B.H.. |
| Homogeneous Redundancy | ..D.. | - | - | R | HR | Data is processed using homogeneous redundant channels; detected faults in data of one channel cause processing to switch to another channel. | IC.Y....M........... |
| Heterogeneous Redundancy | ..D.. | - | - | R | HR | Data is processed using heterogeneous redundant channels detected faults in data of one channel cause processing to switch to another channel. | IC.Y....M........... |
| Data Integrity Sampling | ..D.. | HR | HR | R | R | The integrity of subsets of data is periodically checked, in accordance with a given selection criteria (eg. random, critical records). | IC..O.....L......... |
| Sanity / Reasonability Checks | V.D.. | R | R | HR | HR | Dedicated processing implemented to check that data is within reasonable tolerances and / or logically / semantically consistent (e.g., range checks, date checks, record counts, record sizes, special values - NaN). | I...O.............. |
| Data Correlation | ..D.. | R | R | HR | HR | Data from a number of sources exists to permit a cross-correlation of the data supplied from one source (the master) with other sources. | ICNY............... |

| Technique | Data Types | DSAL | | | | Notes | Data Property |
|---|---|---|---|---|---|---|---|
| | | 1 | 2 | 3 | 4 | | |
| Data Partitioning | ..D.. | R | R | HR | HR | To separate data that is managed differently, creating independence so that a whole data set does not require validation after a change. | ...............B.... |
| Syntax Checks | VID.. | R | R | HR | HR | Semantic checking of data values and sequences based on defined rule sets. | I.N.O.............. |
| Feedback testing | ..D.. | HR | HR | R | R | To check output data by comparing it with the input source. | IC.Y...T.V.......... |
| Information Redundancy | ..D.. | HR | HR | R | R | Additional redundant information is supplied from diverse sources. The validity of the data coming from the diverse sources can be checked against each other. | IC................ |
| Reverse Translation | ..D.. | - | R | HR | HR | To verify data output of a process is correct, by attempting to create the source data from the output data and comparing this with the original source. | IC.Y...T............ |
| Meta-data | ..DP. | - | R | HR | HR | Auditable data are sent with the data that is about the data (e.g., source, issue state, expiry date). | ..N....T.V..PQ....E. |

### 6.6.2.4 Data Design

| Technique | Data Types | DSAL | | | | Notes | Data Property |
|---|---|---|---|---|---|---|---|
| | | 1 | 2 | 3 | 4 | | |
| Governance Model | VI..J | R | R | HR | HR | A governance model is established that defines, e.g., data ownership, processing roles and responsibilities, processing authorisations and permissions. | I....A.T......U.S..D |
| Data Process Definition | VIDPJ | - | R | HR | HR | Documented and agreed process definitions for how data is handled. | .......T......U..... |
| Data Flow Diagram | VIDPJ | HR | HR | HR | HR | To describe the data flow in a diagrammatic form. | ..............U..... |
| Data Model | VIDPJ | HR | HR | HR | HR | To articulate how data is organised. | ..N.O.............. |
| Client Sign-Off | VI.PJ | R | R | HR | HR | Agreement from the client that the data is appropriate. | .......R..V.......... |

| Technique | Data Types | DSAL | | | | Notes | Data Property |
|---|---|---|---|---|---|---|---|
| | | 1 | 2 | 3 | 4 | | |
| Data Quality Correction Mechanisms | ...P. | – | R | HR | HR | A process, strategy and tooling for data that breaches a given data quality criteria. | IC.Y............... |
| Configuration Management | VIDPJ | HR | HR | HR | HR | The recording of the production of every version of every "significant" deliverable and of every relationship between versions of the different deliverable. | .......T.........H.. |
| Data Dictionary | VIDPJ | HR | HR | HR | HR | A collection of descriptions of the data objects or items in a data model for the benefit of data users. | ..N.O.R....F........ |
| Formal Methods | ..D.. | – | R | R | HR | To specify data (or data formats) in a formal, mathematical manner. | .CN.O..T....PQ...... |
| Update Comparison | VIDPJ | – | R | R | HR | Updated data is compared to its previous version (e.g., so the list of changed elements can be compared with a supplier-generated list). | .......T.........H.. |

## 6.6.2.5   Data Implementation

| Technique | Data Types | DSAL | | | | Notes | Data Property |
|---|---|---|---|---|---|---|---|
| | | 1 | 2 | 3 | 4 | | |
| Review / Inspection | VIDPJ | HR | HR | HR | HR | Manual review / inspection of data possibly involving data visualisation tools. | IC..O.....L......... |
| Statistics-Based Sampling | VIDPJ | – | R | HR | HR | More appropriate for real-time large and / or volume data. Could be manual selection, a form of random selection or comparison against statistical norms. | I.NY.A.............. |
| Ground-Truth Check | VIDPJ | R | R | HR | HR | Inspection against physical measurements (e.g., lengths, positions, heights) taken in the real world. | ICN..AR..V.F........ |
| Auditing | VIDPJ | R | R | HR | HR | A period of comprehensive internal and external testing of the data quality process. | ICNYO....V.......... |
| Tracing | VIDPJ | – | R | HR | HR | Ability to trace data from source across multiple participants in the data supply chain. | .......T.V.......... |

| Technique | Data Types | DSAL | | | | Notes | Data Property |
|---|---|---|---|---|---|---|---|
| | | 1 | 2 | 3 | 4 | | |
| Defined Verification Frequency | VIDPJ | - | R | HR | HR | Data should contain an indicator of how often it should be revalidated against other (e.g., real world) source. | .........V.........E. |
| Defined Data Lifetime(s) | VIDPJ | R | R | HR | HR | Information showing when data validity expires. | .................E. |
| Data Quality Trend Analysis | VIDPJ | - | - | R | HR | Checking that a dataset is consistent with a model of the expected data behaviour (e.g., vibration data increases over time). | IC.Y.....V.F......... |
| Authorisation | VIDPJ | R | R | HR | HR | A security model is established to control who is authorised to create, view, edit, delete the data. | ..............UBS..D |
| Authentication | VIDPJ | R | R | HR | HR | Data is authenticated to validate its provenance. | .......T.V.......... |
| Defined Confidence / Trust Levels | VIDPJ | R | R | HR | HR | Criteria are established to provide an objective measurement of the confidence or trust in a given dataset. | IC.Y.....V.F......... |
| Independent Check | VIDPJ | - | - | R | HR | A separate person or system is used to check the data independently. | I.........V.......... |

### 6.6.2.6 Data Migration

| Technique | Data Types | DSAL | | | | Notes | Data Property |
|---|---|---|---|---|---|---|---|
| | | 1 | 2 | 3 | 4 | | |
| Manual Load | ..D.. | R | R | - | - | Data is entered into the system manually relying on human validation and verification. | ICNYOA.....F........ |
| Dedicated Translation and Loading Platform | ..D.. | - | R | HR | HR | For example, using mature enterprise migration Commercial Off-The-Shelf (COTS) products. | ICNYOA.T...F........ |
| Existing / Established System Transfer | ..D.. | - | R | HR | HR | Use of an existing / established proven transfer mechanism. | ICNYOA.T...F........ |
| Client Supervision | VIDPJ | - | R | HR | HR | The client provides independent supervision of activities checking processes, inputs and outputs at agreed points. | ICNYOA.....F........ |

| Technique | Data Types | DSAL | | | | Notes | Data Property |
|---|---|---|---|---|---|---|---|
| | | 1 | 2 | 3 | 4 | | |
| Client Sign-Off | VIDPJ | - | R | HR | HR | Formal acceptance of the migrated datasets in the target system. | ICNYOA.....F........ |
| Incremental switch-over | ..D.. | - | R | HR | HR | Users are incrementally switched over to the new system rather than as a "big bang". | ICNYOA.....F........ |
| Parallel Load With Existing System | ..D.. | - | R | HR | HR | Parallel running of the new system alongside the existing system with data crosschecks between the two systems. | ICNYOA.....F........ |
| Shadowing | ..D.. | - | R | HR | HR | Parallel running of the new system alongside the existing system, only data from the existing system is used operationally, with an experienced user crosschecking between the two systems. | ICNYOA.....F........ |
| End to End Import-Export Verification | ..D.. | - | R | HR | HR | Data is traced and verified at all stages through the entire end to end migration process. | ICNYOA.T...F........ |

### 6.6.2.7  Data Checking

| Technique | Data Types | DSAL | | | | Notes | Data Property |
|---|---|---|---|---|---|---|---|
| | | 1 | 2 | 3 | 4 | | |
| Limited / Pre-Operational Deployment | .IDP. | - | R | HR | HR | A period of monitored operation in a specially chosen environment. | ICN..A.....F........ |
| Client Sign-Off of Data | VI.PJ | - | R | HR | HR | Agreement from the client that the data is appropriate. | ......R..V.......... |
| Non-Critical Trialling | ..D.. | - | R | HR | HR | Monitored operation in an operational, but non-critical, environment. | ......A.....F........ |
| Beta Testing | V.... | - | R | HR | HR | Testing with a small group of specially chosen users. | ICN..A.....F........ |
| Parallel Running | .IDP. | - | R | HR | HR | Running two systems in parallel and crosschecking between them. | ICN..A..M..FP....... |
| Widespread Distribution to User Community | .IDP. | - | R | HR | HR | Large-scale distribution to all users. | ICN.OAR.M.LFP....... |

### 6.6.2.8 Test Data

| Technique | Data Types | DSAL | | | | Notes | Data Property |
|-----------|-----------|---|---|---|---|-------|---------------|
| | | 1 | 2 | 3 | 4 | | |
| Using Informal / ad-hoc means | V.... | R | R | - | - | Data is generated by simple means (e.g, spreadsheets, scripts, basic assumptions). There is no formal checking or review of the method of generation. | ICNY.A.....F........ |
| Using Testbed | V.... | - | R | HR | HR | A dedicated testbed is a good way to produce test data. It may require configuration and tailoring for the particular application, and this configuration should be managed. | ICNY.A.....F........ |
| Using Simulator | V.... | - | R | HR | HR | Simulators (software or hardware) may be able to produce very good test data, obviously depending on how close and detailed a simulation they can achieve. | ICNYOAR....F........ |
| Using Prototype | V.... | - | R | HR | HR | Prototypes are often a good way of generating test data for the real system. However they may not produce data with the appropriate range, accuracy or precision. | ICNY.A.....F........ |
| Using Manual Means | V.... | R | R | - | - | Simple test data can be produced by manual means, although this may be prone to human error. | ICNY.A.....F........ |
| Using Dedicated Platform | V.... | - | R | HR | HR | For complex and critical systems a dedicated test platform is required which can produce realistic test data for all interfaces and inputs. | ICNY.AR..V.FPQ...... |
| Using Existing / Established System | V.... | - | R | HR | HR | Where a new system replaces an old one, then data can often be extracted from the old system to test the new one. Data formats may change so translation may be required. | ICNYOAR.MV.FPQU..... |
| Using Initial Runs of New System | V.... | R | R | R | R | This method is often used where the system is breaking new ground and there is no prototype or legacy system to produce test data. Initial operations may differ from eventual usage, so test data must evolve. | ICNYOAR.MV.FPQ...... |

| Technique | Data Types | DSAL | | | | Notes | Data Property |
|---|---|---|---|---|---|---|---|
| | | 1 | 2 | 3 | 4 | | |
| Derived from Real Data | V.... | R | R | HR | HR | Where real data is available this is usually a good basis for generating test data (e.g., by modification to increase the test space coverage). | ICNY.A.....F........ |
| Statistical Profiling Post-Production | V.... | - | - | R | HR | If a statistical analysis of the data can be produced then greater confidence in the quality of the test data can be obtained. | ICNY.A...V.F........ |
| Produced by Client | V.... | R | R | R | HR | Ideally the client is involved in producing or at least checking the test data. | ICNY.A...V.F........ |
| Client Sign-Off | V.... | R | R | HR | HR | Where possible, the client should formally agree and sign-off the test data as appropriate. | ICNY.A...V.F........ |
| Error Seeding | V.... | R | R | HR | HR | This is where errors are deliberately inserted into the dataset to demonstrate the effectiveness of data validation. | ICNYOAR.MV.F........ |
| Data Reuse | V.... | R | R | HR | HR | Reusing data for one project that was created and thoroughly assured for another project. This can be effective but the read-across should be established. | ICNY.A.....F........ |
| Feedback testing | V.... | R | R | R | R | To check output data by comparing it with the input source. | ICNY.A.....F........ |

## 6.6.2.9   Media – Paper

| Technique | Data Types | DSAL | | | | Notes | Data Property |
|---|---|---|---|---|---|---|---|
| | | 1 | 2 | 3 | 4 | | |
| Photographic Copies | VIDPJ | R | R | HR | HR | Photocopy and store separately. | .C.............B.H.. |
| Scan to Electronic Format | VIDPJ | R | R | HR | HR | Retain both paper and electronic copies. | .C.............B.H.. |
| Copies Held at Different Locations | VIDPJ | - | R | HR | HR | Meaning of "different" depends on data criticality and similarity of location-based risks. | .........VL....B.... |
| Limited Access | VIDPJ | - | R | HR | HR | Control (e.g., by procedure) who can access the data. | ...............U..... |
| Secure Storage | VIDPJ | - | R | HR | HR | Physical measures to prevent unauthorised access. | ...............U..... |

| Technique | Data Types | DSAL | | | | Notes | Data Property |
|---|---|---|---|---|---|---|---|
| | | 1 | 2 | 3 | 4 | | |
| Manual Inspection | VIDPJ | - | R | HR | HR | Used to check data when generated and periodically thereafter. | IC................. |
| Suitable Physical Environment | VIDPJ | - | R | HR | HR | For example, prevent water ingress, control temperature. | I.........L.......E. |
| Defined Handling Procedures | VIDPJ | - | R | HR | HR | To ensure that changes to the data can be attributed. | I.............UB.H.. |
| Repair / Restoration Programme | VIDPJ | - | - | R | HR | To protect against degradation and to ensure availability. | I.........L........ |
| Indexing / Cataloguing | VIDPJ | R | R | HR | HR | To support efficient accessibility. | .........L......... |
| Lifetime Planning | VIDPJ | - | - | R | HR | For example, to avoid gradual quality reduction by repeatedly "copying a copy". | .................ED |

## 6.6.2.10 Media – Electronic

| Technique | Data Types | DSAL | | | | Notes | Data Property |
|---|---|---|---|---|---|---|---|
| | | 1 | 2 | 3 | 4 | | |
| Regular Refresh / Rewrite | VIDPJ | R | R | HR | HR | Of magnetic media or flash memory. | I.................E. |
| Suitable Physical Environment | VIDPJ | R | R | HR | HR | Store media in a clean, low-humidity environment at a steady temperature, cool but not cold. | I.........L.......E. |
| Copies at Different Locations | VIDPJ | R | R | HR | HR | Physically separate to cover natural disasters, accidental or malicious damage. | .........VL....B.... |
| Backups / Duplication | VIDPJ | R | R | HR | HR | Backups are essential. Frequency of backup depends on rate of change. The number of generations to keep relates to the impact of data loss. | .........L......... |
| Sample Restores | VIDPJ | R | R | HR | HR | Sample restores should be performed at intervals to ensure that the backups are readable and retrievable. | .........L......... |
| Multiple Copies | VIDPJ | R | R | HR | HR | At least two backups should be kept, preferably in diverse formats. | .........VL........ |
| Copy to Latest Media Format | VIDPJ | - | R | HR | HR | Anticipate obsolescence and plan a smooth transition to new technologies. | .........L......... |

| Technique | Data Types | DSAL | | | | Notes | Data Property |
|---|---|---|---|---|---|---|---|
| | | 1 | 2 | 3 | 4 | | |
| Media Physically Secured | VIDPJ | - | R | HR | HR | Access to, and removal of, media should be controlled by suitable procedures. Access permissions should be reviewed at intervals. | .......T......U..H.. |
| Resilient / Redundant Format | VIDPJ | - | - | R | HR | This may involve less use of compression, use of error detection and correction protocols, and (at the highest level) two or more redundant data servers. | IC........L......... |
| Long-Lifetime Format | VIDPJ | - | - | R | HR | The best formats should be adopted where available. | ..........L.......E. |
| Easily Translatable / Convertible Format | VIDPJ | - | - | R | HR | Adopt widely-used, well-documented, general-purpose formats in preference to specialist proprietary formats. | ....O.....L......... |
| Copy to Cloud Storage | VIDPJ | - | R | HR | HR | Must specify whether a private cloud or a public cloud shall be used. Cloud storage may not be suitable for highly confidential data. | ..........L......... |
| Copy to Archiving Organisation | VIDPJ | - | R | HR | HR | Consider the required level of data integrity and confidentiality; also, the integrity and long-term viability of the archiving organisation, and plans in case it ceases to function. | ..........L......... |

### 6.6.2.11 Recording the Data Safety Risk Mitigation

The Data Safety Management Plan can be used to document:

- The tables of mitigation measures (or methods and approaches) used for the system, context, and planned implementation under assessment;

- Any specific mitigation measures identified for the system and their source / justification;

- Planned compliance with the tables; and

- Confirmation that the mitigation measures are sufficiently complete and consistent.

The overall safety justification for the given project/service/operational context must then provide evidence of compliance against the plan.

### 6.6.3 Tool Assurance

Tools in this context are considered anything that automates all or part of a process, for example, data creation or data transformation. Test tools are also included (i.e., the term is not limited to parts of an operational system).

Tools can impact data safety in different ways, depending on both their function and how they are to be used. For tools to be considered fit for purpose it is necessary to show that the tool meets its requirements in the context in which it is to be used. The activity to ensure a tool is fit for purpose is usually called "tool qualification".

The first step is to define the purpose for which the tool is required to be fit. Once that is done, and the tool's requirements are specified, there are three main strategies available for qualification:

- Use evidence of a previous certification of the tool by a trusted third party (unlikely to be available in most industry sectors);

- Base tool qualification on the practices used when designing and developing the tool (only practical for tools developed within the organisation); and

- Use one of the available industry-specific guidance documents that admit Commercial-Off-The-Shelf solutions, e.g., EUROCAE Document ED-215 (RTCA/DO-330) [9].

There is an alternative, risk-based and perhaps more practical, approach. This involves assessing the potential risks presented by use of the tool and providing assurance that these risks are adequately managed. The method proceeds as follows:

- Draft a procedure for the use of the tool to achieve the stated purpose;

- Identify threats to data safety associated with using the tool;

- Specify adequate mitigations for each identified threat;

- Augment and formally issue the tool requirements and the usage procedure to implement the specified mitigations;

- Demonstrate that the tool and its mitigations perform as expected; and

- Provide a compelling assurance argument based on the previous steps and any other evidence that will improve confidence; for example:

  - reputation of the supplier;

  - configuration management of the tool, its settings and its documentation;

  - competence of the tool user; and

  - checks that are made on the tool's output.

### 6.6.4 Test Data

The generation of suitable test data is critical to verification of a safety system. The test data must include both representative "normal" values based on intended usage and also values which push at, and beyond, normal use to provoke hazards that the system might produce. This latter type of test data is particularly hard to generate: generally it must be credible, yet it must stress the system to react in a way that the preservation of safety properties can be assessed.

In general, all the properties of the test data should be considered and an assessment made as to whether breaking a property (e.g., introducing corrupt or late data) would cause a problem to the system. If it does, then specific test data should be produced to facilitate testing of this potential problem.

Some suggestions for test data for safety-related systems are:

- Use of values on or around boundaries;

- Use of extreme values, way beyond what could be reasonably expected;

- Use of typical "everyday" values / sets;

- Some realistic but unexpected values;

- Try combinations of data values or items that are problematic together (e.g., inconsistent);

- If possible, use some values known to have caused problems in the past;

- Where appropriate, use values related to timing, rollover or date boundaries;

- Where possible, use white-box values (i.e., those derived from an understanding of the system);

- Use a set of values with drift or bias over time;

- Use data sets with particular statistical properties (e.g., distribution, patterns etc.);

- Use data which has incorrect formatting, ordering, or out of sequence, etc.; and

- Try data which has repeated sets of values or pseudo-random characteristics.

Typically very complex test data is derived from recorded live feeds of real data flows. While this data can be extremely useful for regression purposes, it should be recognised that it is unlikely to contain many outlying or boundary data items. Therefore it may need to be modified to test any hazardous situations; this modification can be difficult and may require sophisticated tools to both ensure correct properties and injection of the intended faults (for instance to introduce a statistical bias to the data).

Simulator / emulator derived values can be useful, but again the issue is how realistic the values are: often the accuracy, resolution or timing of simulated values may be different to real data.

Coverage with test data is something to consider. Sometimes the same data set is used for multiple test scenarios, when in fact it is not stressing all of them to the same degree. Test data coverage can be collected over requirements, code or design, but it is important not to forget hazards: coverage of the hazards and mitigations identified in the hazard log is a key aim.

In general some measure of the quality and suitability of the test data can be useful. This could be based on statistical properties, coverage of hazards or coverage of requirements.

Test data must show continued relevance, through systems evolution and over time. It is good practice to build up extensive regression suites containing coverage of all detected problems to date.

### 6.6.5 Interfaces with Existing Assessments

#### 6.6.5.1 Data and Software

Although most people feel they have an intuitive understanding of the difference between software and data, upon closer examination the boundary is not always as clear as it may first appear.

Consider, for example, Java bytecode, which is operated on by a Java Virtual Machine. From one perspective, it could be argued that the Java bytecode is simply data. By extension, it could also be argued that the Java source code is also just data. This type of argument can be extended to suggest that any software can, at least from one viewpoint, be considered as data. Conversely, think about the data used in a 3D printer, perhaps to produce a part for an aircraft. This data could be viewed as a program for the printer; that is, it could potentially be viewed as software. This type of argument can easily be extended across a range of situations, especially those relating to configuration data.

Whilst they are interesting, and potentially important, these philosophical considerations should not detract from the practical issue: there are some aspects of data (using the term in a generic sense) that are often not explicitly addressed in standards. These are a consequence of features that are more readily apparent in data than in software. Examples include:

- It is easy for data to be reused in a range of contexts and despite appearances it is not trivial to translate an assurance argument that the data is fit-for-purpose from one context to another.

- It is not always clear who owns or is responsible for data, especially when data is shared and processed amongst a collection of disparate systems.

- There is often a default value for data. Whilst this can make systems easier to use and hence more productive it can be difficult to identify a default value that is appropriate for all circumstances.

- It can be easy to change data. In some circumstances this can give rise to a temptation to make uncontrolled and potentially untested changes.

In summary, data and software are closely related and, as such, need to be considered together in system engineering activities, including system safety analyses. However, data and software emphasise different facets of risk and they are susceptible to different mitigation approaches; this means there is also a need to adopt a data-focussed perspective. It also means that assurance levels related to software cannot be mapped directly to Data Safety Assurance Levels.

### 6.6.5.2   Data Safety and Security

When generating high-level processes and techniques to manage the risks posed by data, it is worthwhile understanding the difference between the safety risks posed by accidental failure to preserve data properties and the security risks posed by actors maliciously undermining the properties of data.

The relationship between safety and security, as engineering concepts, can be summarised by their relationships to cultural, developmental and aspirational properties of systems development.

Culturally, embedding both safety and security into an organisation is seen as a key strategic goal for creating systems that are both safe and secure. Developmentally, safety and security are quality factors, generating transverse requirements that impact the entire system. Most importantly, at the aspirational level, both safety and security have the common goal of preventing harm from accidental and malicious interventions respectively.

For an organisation aiming to create systems that are both safe and secure, these connections can be both a benefit and a burden. The shared goal of preventing harm means that both quality factors seek to identify routes to harm through analysis of the system being developed. This can result in shared processes and tools, which in turn can save time and money during systems development. However, safety and security interact in a more volatile way at the functional level. Security failings can undermine the safety case for a system and, conversely, safety requirements can prevent the implementation of standard security solutions. For example, the German government published a report in 2014 into a fire at a steel works caused by a cyber attack that resulted in the control system being placed into an unsafe state and the safety system being unable to intervene (Section 3.3.1 of [5] - in German). In addition, "fail-safe" states can often leave a system with exposed security vulnerabilities.

These links between safety and security infer that there are connections between the sub-categories of data safety and information security: both attempt to take a data-centric view of the system of interest in order to improve the associated quality factor; and both attempt to prevent harm through the preservation of the properties of data within that system.

In the security domain, the three key properties of data considered are confidentiality, integrity, and availability. Confidentiality, (the failure of which is termed "Information Disclosure" in the Microsoft Security Model, [30]) is typically not a safety concern as, without malicious intent, information sharing is not inherently unsafe. However, when considering systems where confidentiality is an important property, the interaction between data safety and security cannot be trivially resolved. For example, accidental disclosure of information can form part of a causal chain which leads to harm from a malicious actor.

Data integrity is a critical property for both domains. The Microsoft Security Model describes malicious removal of the property of integrity as "tampering". Whether by accident or through malicious intent, the potential harm from loss of data integrity can be disastrous to a safety-critical system, from the values of drug dosages to control system parameters.

Data availability is also important to both domains. Loss of availability, or "denial of service" in the Microsoft Security Model, is another property that can be lost accidentally or through malicious intervention. Loss of availability prevents systems from functioning properly and can result in undefined behaviour if not mitigated by design.

# 7 Worked Example – Healthcare

> *There is a forest of data and we need to create a path through.*
> **Tom Adams**

## 7.1 Purpose

This section provides a worked example of applying the Data Safety guidance to a hypothetical system in the Healthcare sector. Although some aspects of the example have been simplified, it is intended to be sufficiently realistic to allow key features of the guidance to be illustrated.

## 7.2 Establish Context

### 7.2.1 Background

A Manufacturer is building a new integrated health and social care system to support holistic care for community health services. The system supports clinical workflows for aspects such as referrals, tracking clinical encounters, appointment scheduling, outcome measures through to letter and report generation. The system follows a typical development lifecycle as a series of phases: business modelling; requirements; analysis and design; implementation; test; and deployment.

The system is being targeted to meet the requirements of a Health Organisation who are procuring a solution to help their clinicians maintain a high level of quality of care in the face of increasing volumes of patients and pressure to reduce staff costs.

*From a data perspective, fundamental to fulfilling these requirements is establishing the context within which the use of data in system development, enhancement, introduction, integration or operation is occurring. This should establish the risk appetite: essentially, how much effort is devoted to making data risks as low as practicable. In turn, this will inform the nature and scope of assessments that are conducted during system development and, furthermore, its introduction into operational service. To meet the 'Establish Context' objectives set out in the guidance the following activities are recommended:*

- *Describe the organisational context;*

- *Describe the system context;*

- *Plan the assessment; and*

- *Identify Data Artefacts.*

The Manufacturer decides to use an Organisational Data Risk (ODR) Assessment form to understand in broad terms the level of risk it will have to manage in developing and supporting this system. This will allow the Manufacturer to **Describe the organisational context** and **Describe the system context.**

**System Manufacturer**

### 7.2.2  Organisational Data Risk (ODR) Assessment

The following list identifies the questions in the ODR, along with the assessments for the Manufacturer's system.

| Q1: | How severe could an accident be that is related to the data? Could it be caused directly by the data? |
|---|---|

Failings in the system could give rise to non-optimal treatment plans for a patient that might delay detection of a more serious condition or prolong the recovery for a known condition. The system is not solely relied upon however, and there are other people and systems in play involved in checking data. On balance, this question is assessed as: **1c**; Score **4**.

| Q2: | What would be the impact on the organisation, client or public if an accident occurred related to the data? |
|---|---|

Unfortunately, accidents in the health domain are relatively frequent. There are many injuries and deaths attributed to medical errors but these are largely tolerated by the public and grievances are usually settled through financial settlements through the courts. The Manufacturer believes through their contractual arrangements that the Health Organisation would be liable for any claims ·even if it was attributed to an error in the Manufacturer's system's handling of data. On balance, this question is assessed as: **2b**; Score **2**.

| Q3: | How much responsibility does this organisation have for data safety? |
|---|---|

The Manufacturer is responsible for building the system in compliance with SCCI0129 and so is responsible for executing the associated safety management system to manage risk. The Manufacturer however plans to sell the product with a condition of use that places end responsibility for patient safety on the client. On balance, this is question is assessed as: **3b**; Score **2**.

| Q4: | What legal and regulatory environment will this work be subject to? |
|---|---|

The work will be contracted under UK law and subject to the Standardisation Committee for Care Information (SCCI) standards for Health IT Systems. However, there is no special regulator who is currently empowered to intervene in the delivery of healthcare systems, i.e., the standards are not currently enforced through law. The Health Organisation however will make compliance with the standards a contractual requirement. On balance, this question is assessed as: **4c**; Score **4**.

| Q5: | How mature is this organisation regarding data safety? |
| --- | --- |

The Manufacturer has a good understanding of data as a source of safety risk. Many of their systems are data intensive to support clinical decision making. There is good support and funding for the identification and resolution of data-related risks. On balance, this question is assessed as: **5b**; Score **2**.

| Q6: | How widely used is the data and who by? |
| --- | --- |

The data will be used in multiple clinical settings and by many clinicians and other support staff. There are several data supply chains and public web access to data. On balance, this question is assessed as: **6c**; Score **4**.

| Q7: | What is the scale, sophistication and complexity of the data and its manipulation? |
| --- | --- |

The data is complex and although transmitted through industry standard data structures, these require knowledge of the associated abstract clinical data model. Some data manipulations are required to map between different encodings for data held in the various heterogeneous systems. Some legacy systems transfer data in unstructured format. On balance, this question is assessed as: **7c**; Score **4**.

| Q8: | How well defined and understood are the boundaries and interfaces for this data scenario? |
| --- | --- |

The boundaries of the supply are well understood and although the interfaces are complex and mixed formats, these will be defined and agreed formally through Interface Control Documents. Most of the integrating systems are established COTS based systems but some of the legacy systems still need to be investigated and working assumptions have been by the Manufacturer. On balance, this question is assessed as: **8c**; Score **4**.

The final score is **26**, which corresponds to **ODR2**. The Manufacturer therefore concludes that there is low to medium risk that loss of properties of data in the system can contribute to or give rise to harm. The Manufacturer has an internal policy for engagements based on the ODR level that dictates how the organisation shall *Plan the assessment*; this policy dictates the amount of proportional effort it needs to spend on safety data management and the level of rigour to be employed. In this case, the policy dictates, amongst other requirements, that a separate section covering Safety Data Management is required in its Clinical Risk Management Plan.

The Manufacturer aims now to *Identify Data Artefacts* that are potential sources of safety hazards. The Manufacturer also knows that the safety dependency of data is dictated by the context in which it is used so it now develops an understanding of when in its process lifecycle the data will be used and relied upon. The Manufacturer plans to build an early prototype to show to clients to help elicit requirements definition. To support this, the Manufacturer plans to create a test data set that comprises a typical range of scenarios that the system will encounter. This form of data is identified as **Verification Data**. The system also needs to be configured to support deploying Health Organisations' policies. This data is **Infrastructure Data** and, for the prototyping phase, the Manufacturer plans to use largely default values.

In later phases when the system functionality is specified and the system is being built, the Manufacturer plans to create a test data set that will be key to demonstrating the correct functioning of the system and hence acceptance by the deploying Health Organisation. This still involves the use of **Verification Data** and **Infrastructure Data** but there will be far greater dependency on these data sets than the prototyping case. The Manufacturer therefore documents the planned use of each of the data types during the entire delivery lifecycle in the Clinical Safety Management Plan.

**Health Organisation**

*The procuring Health Organisation will have a different perspective of the IT system that they will deploy into their organisation. They will already have many integrated systems in live operation and as part of establishing the context for the system's deployment they will need to consider many different types of data sets:*

- *Infrastructure Data: how the system will be configured in the specific environment;*

- *Verification Data: the test data sets to be used to support certain deployments such as integration testing and training; and*

- *Dynamic Data: the data entered or fed into the system and the data presented to the user, generated in the form of reports or data passed to other systems.*

The Health Organisation decides to complete an Organisational Data Risk Assessments so it can *Describe the organisational context* and *Describe the system context*. The scoring is similar to the Manufacturer with the following notable differences:

| Q2: | What would be the impact on the organisation, client or public if an accident occurred related to the data? |
|---|---|

The Health Organisation would bear the brunt of any publicity and litigation in the event of an accident and so assess this question as **2c**, Score **4**.

| Q3: | How much responsibility does this organisation have for data safety? |
|---|---|

The Health Organisation is responsible for deploying systems in compliance with SCCI0160 and is ultimately responsible for patient safety. The Health Organisation assesses this question as **3e**, Score **12**.

| Q5: | How mature is this organisation regarding data safety? |
|---|---|

The Health Organisation has only recently acquired the expertise to apply SCCI0160 and is still developing its capability. Safety Data Management is new to the organisation and it anticipates some resistance from senior management after the expenditure incurred in rolling out a SCCI0160 compliant safety management system. This question is assessed as **5d**, Score **7**.

The resulting score for the Health Organisation is **43**. This is **ODR3**; medium to high risk. The Health Organisation aims to *Plan the assessment* through a Clinical Risk Management Plan. This plan defines the organisation and system context in more detail and lays out the planned activities for identifying, evaluating and treating data safety related risks.

As with the Manufacturer, the Health Organisation needs to *Identify Data Artefacts* that are potential sources of safety hazards and understand the context of their use in its lifecycle. Post acceptance, the procuring Health Organisation plans to run a series of user training sessions for clinicians. Once users are trained, the system will be integrated into live operations. The Health Organisations identifies the Infrastructure, Verification and Dynamic Data Types to be used during these phases. The Health Organisation also realises that the system will form part of a data supply chain as a number of external organisations and departments within their own organisation engage in the procurement and use of safety-related data. For example, it will receive referral data from a number of other General Practitioner (GP) systems, it will receive outcome measures from hospitals and clinical data acquired from remote workers visiting patients in the community and also from the patients themselves using the system's online portal. The system also produces data for other external systems such as electronic prescriptions for pharmacies.

The Health Organisation sees that by using the new system it will become a Commissioning User as it will require and be a **Consumer** of data from GP's systems, hospital systems, systems used by remote worker in the community and the system's portal capturing data entered by the patients themselves, each of these acting as **Data Provisioners**. Those health care professional (and the patient themselves) gathering patient data through physical inspections and measurement are the **Data Acquirers**.

The Health Organisation defines the data supply chain relevant to the system including the roles and interfaces involved in its Clinical Risk Management Plan. This will therefore show where there are dependencies on **Dynamic Data** used and produced by the system.

Questions the Health Organisation will need to address when establishing the context are:

- Have all the dependent interfaces been identified?

- Have the roles of Commissioning User / Data Provider / Data Acquirer been established and acknowledged?

- What 'service levels' or contracts exist for the delivery of the data?

- What level of assurance do Data Providers/Data Acquirers provide for their data?

## 7.3   Risk Identification

The Manufacturer aims to carry out the following activities to meet the guidance objectives for the Risk Identification phase.

**System Manufacturer**

- Review the general, historical perspective;

- Conduct a top-down approach;

- Conduct a bottom-up approach; and

- Update planning documents.

Before embarking on any hazard analysis, the Manufacturer ensures that stakeholders *Review the general, historical perspective*. This takes the form of a refresh briefing to raise awareness of issues that are specific to data such as ageing, biasing and defaults.

The Manufacturer of the Health IT System decides that during the prototyping phase there is little safety dependency of the test and configuration data sets as no clinical decisions will be made based on their content; the data is simply being used to support the elaboration of requirements.

In later phases however, when the system functionality is specified and the system is being built, the Manufacturer will want to create a test data set that will be instrumental in demonstrating the correct functioning of the system. This still involves the use of **Verification Data** and **Infrastructure Data** but there is far greater dependency on these data sets than the previous case. For example, if the verification or configuration data is not sufficiently diverse or insufficiently models real world scenarios, it is possible that erroneous and unsafe functional behaviour is present in the system during live operation despite this system having passed factory and site acceptance testing.

To analyse the risks in more detail the Manufacturer uses a *Top-down approach* and a *Bottom-up approach*. In the first approach it considers each of the system functions (such as clinical screens) and analyses where there is a dependency on data and the properties that need to be preserved. In the second, as much of the functionality is driven by data flows in and out of the system, the Manufacturer also looks at specific data flows and assesses the impact if there are loss of properties for the data in those flows.

On completion of the risk identification phase the Manufacturer *Updates planning documents* such as the Clinical Risk Management Plan to reflect the outcome of the analysis.

The Health Organisation will likewise need to conduct risk identification relevant to their deployment context. Hazards arising from data sources that are to be delivered into the new system from existing systems need to be assessed for data risks.

**Health Organisation**

As with the Manufacturer, a briefing to stakeholders to *Review the general, historical perspective* is first conducted to cover generic data safety issues but also to highlight lessons learnt from previous accidents and incidents that have occurred in the Health Organisation itself.

As with the Manufacturer, both a *Top-down approach* and a *Bottom-up approach* is adopted. From the Health Organisation's perspective, one key focus for hazard identification is in the use of **Dynamic Data**, i.e., data that will be delivered into the new system from existing system data sources, and the data presented to the user. For the interactions identified in the supply chain, the Health Organisation needs to consider the risks associated with loss of properties of the data it will receive. Questions the Health Organisation will need to consider and address more formally in the Clinical Risk Management Plan are as follows:

- Which data sets or items being received from other systems have Data Properties (such as timeliness, completeness, consistency, fidelity etc.) that are significant to patient safety?

- What data presented to the user has Data Properties (such as availability, format, resolution, etc.) that are significant to patient safety?

- What existing barriers or mitigations (physical, technical, procedural) exist to reduce the risk of loss of Data Properties?

On completion of the Risk Identification phase the Health Organisation will *Update planning documents* such as the Clinical Risk Management Plan to reflect the outcome of the analysis.

## 7.4   Risk Analysis

In this phase identified hazards are assessed to determine their likelihood and severity. To meet the guidance objectives the following activities are carried out:

- Establish DSALs; and

- Analyse DSALs as part of system safety activities.

The Manufacturer will *Establish DSALs* by considering cases where the use of specific types of data could give rise to hazards. In the first, the prototyping phase, the Manufacturer sees no use of the data that can give rise to credible clinical risk and assessing the DSAL for that data set as **DSAL0**.

**System Manufacturer**

In the second phase of the development lifecycle, where **Verification Data** and **Infrastructure Data** is being used to demonstrate the correct functioning of the system, the Manufacturer considers that loss of any of the Data Properties of **Integrity, Completeness, Consistency, Continuity, Format, Accuracy, Resolution, Timeliness, Availability, Fidelity / Representation, Sequencing, Intended Destination / Usage** of this data could give rise to hazards.

For example, if the verification data set selected is not representative of the eventual diversity experienced in practice (loss of Fidelity / Representation), then it is possible that the system may contain latent software errors that could give rise to harm. However, the Manufacturer acknowledges that the system will be subject to further testing and trials in the clinical setting and so there will be other opportunities to detect errors in the system. Overall:

- The likelihood of the data use gives rise to an accident is **Medium** as other systems and processes are in place that would detect errors; and

- The severity is **Moderate**; failings in the system could give rise to non-optimal treatment plans for a patient that might delay detection of a more serious condition or prolong the recovery for a known condition.

The Manufacturer therefore assesses these data types as **DSAL1** in this particular context of use.

The Manufacturer takes care to *Analyse DSALs as part of system safety activities* by documenting these assessments along with other hardware and software safety considerations in the Clinical Risk Management Plan.

From the Health Organisation's perspective, the main focus for risk assessment is in the use of **Dynamic Data**. For the interactions identified in the supply chain the Health Organisation needs to consider the risks associated with loss of properties of the data it will receive and present to the user. Questions the Health Organisation will need to consider and address more formally in the Clinical Risk Management Plan are as follows:

**Health Organisation**

- How likely is it that there would be a loss of the given Data Property?

- How would such a loss of a Data Property be detected?

- How would such as loss be isolated to prevent further risks of harm?

- What recovery action would be required to resolve the issue to maintain patient safety?

In considering the receipt of outcome measures data received from a clinic or hospital, the Health Organisations considers that it is likely that some credible errors would not be readily detected by their new system; if the hospital system confused a result or there were errors in the precision of data then there would be few chances to catch these once received by the system.

- The Health Organisations assesses the likelihood of this loss of property as **High**; and

- The impact of such errors, although not realistically likely to lead to death, could result in delays to treatment that could result in serious injury and hence **Moderate** impact.

The data received from this data source is therefore classed as **DSAL2** in this particular context of use.

## 7.5   Risk Evaluation and Treatment

The Manufacturer carries out the following activities to meet the guidance objectives:

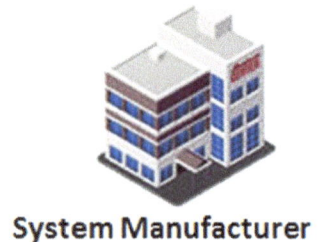

**System Manufacturer**

- Review each risk and either: Avoid; Accept; Transfer; Treat;

- Establish treatment methods for relevant risks

- Implement and verify treatment methods

The Manufacturer decides to *Review each risk and either: Avoid; Accept; Transfer; or Treat the risk*. It decides to **Accept** the **DSAL0** risk but as it has determined there is some, albeit low, risk (**DSAL1**) associated with its use of data at a specific point of its lifecycle, it decides to **Treat** that risk. The Manufacturer evaluates this risk and considers that the risks should be reduced further by taking some reasonably practicable steps.

Likewise, the Health Organisation has identified DSAL2 data and in deciding to Treat the risk, it aims to ensure risks are reduced as low as reasonably practicable.

**Health Organisation**

Having decided that further risk reduction is necessary, the Manufacturer needs to *Establish treatment methods for relevant risks* that are appropriate for DSAL1 data and in doing so demonstrate that reasonably practicable steps have been taken to reduce the risk. The Manufacturer therefore refers to the tables in the Data Safety Guidance document. The Manufacturer then documents in its Clinical Risk Management Plan:

- Planned compliance with the tables;

- The interpretation for the given method/technique (e.g. depth of checking); and

- Justification in the case where a technique is not to be adopted.

For **DSAL1** Verification Data, the tables show that the following are recommended (R) or highly recommended (HR) where loss of data properties **Integrity, Completeness, Consistency, Continuity, Format, Accuracy, Resolution, Timeliness, Availability, Fidelity / Representation, Sequencing, Intended Destination / Usage** can give rise to hazards:

| Ref | Technique | R/HR | System Design |
|-----|-----------|------|---------------|
| SD1 | Syntax Check | R | Semantic checking of data values and sequences based on defined rule. |
| SD2 | Sanity / Reasonability Checks | R | Dedicated processing implemented to check that data is within reasonable tolerances and/or logically/semantically consistent with what the data represents. For example, range checks, date checks, record counts, record sizes, special values (e.g., NaN) etc. |

| Ref | Technique | R/HR | Data Design |
|-----|-----------|------|-------------|
| DD1 | Governance Model | R | A governance model is established that defines aspects such as data ownership, processing roles and responsibilities (who can do what to the data), processing authorisations and permissions (what can be done to the data) etc. |
| DD2 | Data Flow Diagram | HR | To describe the data flow in a diagrammatic form. |
| DD3 | Data Model | HR | To articulate how data is organised. |
| DD4 | Client Sign-Off | R | |
| DD5 | Configuration Management | HR | The recording of the production of every version of every "significant" deliverable and of every relationship between versions of the different deliverable. |
| DD6 | Data Dictionary | HR | A data dictionary is a collection of descriptions of the data objects or items in a data model for the benefit of data users. |

| Ref | Technique | R/HR | Data Implementation |
|-----|-----------|------|---------------------|
| DI1 | Review / Inspection | HR | Manual review / inspection of data possibly involving data visualisation tools. |
| DI2 | Ground-Truth Check | R | Inspection against physical measurements (e.g., lengths, positions, heights) taken in the real world. |
| DI3 | Auditing | R | A period of comprehensive internal and external testing of the data quality process, where Data is verified according to its intended use and definition. |
| DI4 | Authorisation | R | A security model is established to control who is authorised to create, view, edit, delete the data. |
| DI5 | Authentication | R | Data is authenticated to validate its provenance. |
| DI6 | Defined Confidence / Trust Levels | R | Criteria are established to provide an objective measurement of the confidence or trust in a given dataset. |

| Ref | Technique | R/HR | Test Data |
|-----|-----------|------|-----------|
| TD1 | Using Informal / ad-hoc means | R | This is where data is generated by simple spreadsheets, or by simple scripts or programmes. It may also be legacy data or basic assumptions. There is no formal checking or review of the method of generation. |
| TD2 | Using Manual means | R | Simple test data can be produced by manual means, although this may be prone to human error. However manual checking of a sample of test data generated using tools is a useful verification method. |
| TD3 | Using Initial Runs of New System | R | This method is often used where the system is breaking new ground and there is no prototype or legacy system to produce test data. This must be carefully used as initial operations can be very different to eventual usage, and so the test data suite must also evolve. |
| TD4 | Derived from Real Data | R | Where real data is available this is usually a good basis for generating test data (e.g. by modification to increase the test space coverage). However there are potential issues of sampling and coverage, i.e. is the real data a representative sample? |
| TD5 | Produced by Client | R | Ideally the client is involved in producing or at least checking the test data. The client will often know the data intimately and can highlight any issues quickly. |
| TD6 | Client Sign-Off | R | Where possible, the client should formally agree and sign-off the test data as appropriate. This gives the system developer some confidence in the data and also some protection of the data is in fact incorrect or not representative. |

| Ref | Technique | R/HR | Test Data |
|-----|-----------|------|-----------|
| TD7 | Error Seeding | R | This is where errors are deliberately inserted into the dataset to demonstrate the effectiveness of data validation. |
| TD8 | Data Reuse | R | Reusing data for one project that was created and thoroughly assured for another project. This can be effective but the read-across should be established |
| TD9 | Feedback testing | R | To check output data by comparing it. |

| Ref | Technique | R/HR | Data Migration |
|-----|-----------|------|----------------|
| No relevant techniques | | | |

| Ref | Technique | R/HR | Data Checking |
|-----|-----------|------|---------------|
| No relevant techniques | | | |

| Ref | Technique | R/HR | Media – Paper |
|-----|-----------|------|---------------|
| MP1 | Photographic Copies | R | |
| MP2 | Scan to Electronic Format | R | |
| MP3 | Indexing / Cataloguing | R | |

| Ref | Technique | R/HR | Media – Electronic |
|-----|-----------|------|--------------------|
| ME1 | Regular Refresh / Rewrite | R | Of magnetic media or flash memory. Life of a hard disk might be <3 years. Life of a properly-stored DVD might be 5 years. Life of a USB stick might be 10 years. Life of a magnetic tape might be longer if a clean tape drive was used and the tape was stored properly. |
| ME2 | Suitable Physical Environment | R | Store media in a clean, low-humidity environment at a steady temperature, cool but not cold. |
| ME3 | Copies at Different Locations | R | Physically separate to cover natural disasters, accidental or malicious damage. |
| ME4 | Backups / Duplication | R | Backups are essential. Frequency of backup is dependent on the rate of change of the data. The number of generations of backup to be kept should be commensurate with the impact of data loss. |
| ME5 | Sample Restores | R | Sample restores should be performed at intervals to ensure that the backups are readable and retrievable. |

From these tables the Manufacturer decides on a series of activities to implement the recommendations that are applicable to its particular endeavour. These activities are expressed as a series of requirements that can be placed on the Manufacturer's delivery organisation and tracked through to completion.

| Ref | Requirement | Guidance Reference |
|-----|-------------|--------------------|
| R1 | The verification data shall be carefully controlled in the Manufacturer's configuration management system. There shall be a configuration management plan that shall define who has responsibility for the data and who is authorised to create and amend it. | DI4, DI5, DD1, DD5 |
| R2 | The verification data shall be held on an industry standard fileshare that is regularly backed up with copies moved periodically to offsite storage. The Backup / Recovery plans shall include periodic sampling of restores. | ME1, ME2, ME3, ME4, ME5 |
| R3 | The data shall be modelled as a series of patient "journeys" that cover the entire lifecycle of data from first encounter through to archival and deletion of data. The complete set of journeys shall be chosen to exercise all the functionality of the system. The modelling shall include a data dictionary, data flow diagrams and a data model. | DD2, DD3, DD6 |
| R4 | To model data from external systems, the Manufacturer shall use manual data entry and spreadsheet based records to hold the data. | TD1, TD2 |
| R5 | The Manufacturer already has a set of clinical standing data that was used for another system and derived from real data. This data includes data such as encounter codes, clinical terms, consultant names, surgery and hospital addresses etc. and this shall be reused for this system. The Manufacturer's Clinical Safety Officer has reviewed the data and agreed on its suitability for reuse. | TD4, TD8 |
| R6 | Some of the verification data sets shall include errors deliberately inserted to check the effectiveness of data validation. | TD7 |
| R7 | The controlled verification data set shall be subject to review and analysis against defined confidence/trust criteria. Scripts shall be written to check for syntax and semantic consistency of the data and provide a basic sanity check. The scripts themselves shall be validated and verified before use. | SD1, DI1, SD2, DI2, DI6 |
| R8 | The project shall be subject to an internal delivery quality assurance audit. | DI3 |
| R9 | Data loaded from external system into the system and displayed to the user shall be crosschecked against the original source data, using manual spot-checks. | TD9 |
| R10 | The level of rigour employed in verifying all the above requirements shall be commensurate with the DSAL criticality and so an ISO9001 compliant quality management system shall be adopted. | All |

The following guidance recommendations were not adopted by the Manufacturer for the reasons given. Note that some may however become relevant in the future so actions are set, where appropriate, to review the applicability of the recommendation when the given condition is met.

| Ref | Guidance Reference | Justification | Action |
|-----|--------------------|---------------|--------|
| E1 | TD3 | The data will be used before any initial run of the system. | Review when data from initial runs is available. |
| E2 | DD4, TD5, TD6 | There is no contracted client at the moment as the system is a new developed so it will not be possible to get the client to create or signoff data. | Review when contracting with a client. |
| E3 | MP1, MP2, MP3 | There are no paper based resources for this system | No further action. |

Having determined the requirements arising from the data safety analysis the Manufacturer ensures these are included along with other system requirements as part of the overall delivery and operation of the system. It then remains for the Manufacturer to *Implement and verify treatment methods*, that is, as well as defining requirements the Clinical Risk Management Plan needs to ensure activities are in place to verify and evidence that treatments have actually been implemented.

# 8 Conclusions

*The world is one big data problem.*
**Andrew McAfee**

The nature of systems is changing, with the role of data becoming ever more prominent. This means that data and, more specifically, the properties it is required to exhibit have a direct effect on system safety. A number of accidents and incidents have already occurred in which "inappropriate data" was amongst the causal factors.

Raising data to a level where it is considered as a "first class citizen" alongside software and hardware in system safety analyses can help protect against such occurrences. This guidance document has outlined a Data Safety Management Process to facilitate this. The process includes: a set of Data Types and associated Data Properties; a way of establishing the appropriate context (e.g. for system assessments); and methods for identifying, analysing, evaluating and treating risks.

Although it is not a panacea, and it is still under active development, this Data Safety Guidance document provides a means of managing and mitigating the risks associated with the use of data in safety-related systems. As such, the authors hope it will provide a valuable contribution to the development and safe operation of such systems.

# Appendix A   Incidents and Accidents

*Hiding within those mounds of data is knowledge that could change the life of a patient, or change the world*
**Atul Butte**

## A.1 General

The following 'War Stories' describe incidents and accidents in which data is considered to have been a contributory factor. A data perspective has been taken to demonstrate the need for data to be given equal footing alongside software, hardware and human factors. The items described here have been arbitrarily selected; the collection is not intended to be exhaustive.

**Note:** The analysis presented here has no legal standing whatsoever. The purpose of this section is not to discredit, contradict or undermine any existing accident analysis; the aim is simply to view these incidents from a data perspective. Where possible accident reports have been referenced with the role of data highlighted. All references have been taken at face value and not independently verified.

## A.2 A400M, Torque Calibration Parameters

**Summary** On May 9 2015, just minutes into a routine, pre-delivery test flight an Airbus A400M military plane, crashed in Spain, killing four of the six crew. Three of the four engines had become stuck at high power and initially did not respond to the crew's attempts to control the power setting in the normal way. Pilots then succeeded in reducing power only after selecting the thrust levers to idle. The engines subsequently remained stuck in this mode. In an attempt to return to the airport, the aircraft struck powerlines and crashed. An Airbus official after the accident stated that engine control software was incorrectly installed during final assembly of the aircraft.

**Role of Data** Although not confirmed, reports suggest the torque calibration parameters for the engines were wiped during the installation. Hence, this accident demonstrates the importance of the completeness Data Property.

The torque calibration data is needed to measure and interpret information coming back from the A400M's engines, and is crucial for the Electronic Control Units (ECU) that control the aircraft's power systems. Without that sensor data, the ECU automatically shut down the engines, or at least put them into the lowest power settings. The ECUs take the pilot's inputs and make the engines they control respond in the optimum way. The parameter files are used by the ECUs to interpret sensor readings about the turning force generated by each engine - the torque - which is used to make the attached propellers spin. Without the files, the ECUs cannot make sense of this data. On May 20, Airbus warned A400M customers to conduct "specific checks of the Electronic Control Units (ECU) on each of the aircraft's engines".

**Sources**

- BBC News, http://www.bbc.co.uk/news/technology-33078767, accessed on 18 November 2016.

- Reuters, http://www.reuters.com/article/us-airbus-a400m-idUSKBN0OP2AS20150609, accessed on 18 November 2016.

## A.3 Turkish Airlines A330

**Summary** During March 2015, an Airbus A330-303, operated by THY Turkish Airlines, suffered a runway excursion accident upon landing at Kathmandu-Tribhuvan Airport (KTM), Nepal.

Flight TK726 was a regular passenger service from Istanbul-Atatürk International Airport (IST) to Kathmandu, Nepal. The flight was the first international flight to arrive that morning. After descending from cruising altitude, it entered a holding pattern at FL210 at 06:12 hours local time (00:27 UTC) until about 07:00 (01:15 UTC) when it was cleared for a VHF Omnidirectional Range (VOR) / Distance Measuring Equipment (DME) approach to Runway 02.

This approach was abandoned at about the Missed Approach Point at 1DME and the aircraft performed a go around. The aircraft circled and positioned for a second approach to Runway 02. The aircraft touched down to the left of the runway centre line with the left hand main gear off the paved runway surface. It ran onto soft soil and the nose landing gear collapsed. Following the accident the aircraft was written off.

**Role of Data** The aircraft touched down at to the left of the centreline because the Flight Management Guidance System (FMGS) navigation database contained threshold coordinates for a proposed displacement of the Runway 02 threshold. This was later withdrawn through a Notice to Airmen (NOTAM), but had not been updated by the airline in the database. Additionally, the coordinates that were initially published were inaccurate, causing the threshold coordinates to be offset to the left of the actual threshold. This had been noticed and reported by a previous Turkish Airlines flight on March 2. The changes had not been performed by the time TC-JOC landed at Kathmandu.

Among the safety recommendations stated in the accident report were:

- "The operator must ensure that the correct navigation data are uploaded on Flight Management Guidance System";

- "The operator should establish a system of verifying the quality of charts prepared by the service provider";

- "The operator should establish a system of checking the validity of Flight Management System database"; and

- "Civil Aviation Authority of Nepal must ensure that raw aeronautical information/data are provided by the aerodrome authorities taking into account of its accuracy and integrity requirements for aeronautical data as specified by ICAO Annex 15 and its Aeronautical Information Service Manual."

### Sources

- Aircraft Accident Investigation Commission, http://www.tourism.gov.np/uploaded//TURKISH-AIRLINE-Final-Report-finalcopy.pdf, accessed on 18 November 2016.

## A.4 Qantas Boeing 737 Take-Off Performance Data

**Summary** On 1 August 2014 a Qantas Boeing 737-838 aircraft (registered VH-VZR and operated as QF842) commenced take-off from Sydney Airport, New South Wales. The flight was a scheduled passenger service from Sydney to Darwin, Northern Territory.

While the aircraft was climbing to cruise level, a cabin crew member reported hearing a "squeak" during rotation. Suspecting a tailstrike, the flight crew conducted the tailstrike checklist and contacted the operator's maintenance support. With no indication of a tailstrike, they continued to Darwin and landed normally. After landing, the captain noticed some paint was scraped off the protective tailskid. This indicated the aircraft's tail only just contacted the ground during take-off

**Role of Data** The Australian Transport Safety Bureau (ATSB) found the tailstrike was the result of two independent and inadvertent data entry errors in calculating the take-off performance data. As a result, the take-off weight used was 10 tonnes lower than the actual weight. This resulted in the take-off speeds and engine thrust setting calculated and used for the take-off being too low. Hence, when the aircraft was rotated, it overpitched and contacted the runway.

The ATSB also identified that the Qantas procedure for conducting a check of the Vref40 speed could be misinterpreted. This negated the effectiveness of that check as a defence for identifying data entry errors. In this case, uncorrected errors affected the integrity of the data used to calculate take-off parameters.

Sources

- ATSB, http://www.atsb.gov.au/publications/investigation_reports/2014/aair/ao-2014-162.aspx, accessed on 18 November 2016.

## A.5 Qantas Boeing 737 Loading Incident

**Summary** On 9 May 2014 a Qantas Boeing 737 was preparing for departure from Canberra to Perth. There were 150 passengers, 87 of which were primary school children. These children were all seated together at the rear of the cabin. All had been assigned an 'adult weight' of 87 kg.

During take-off the aircraft appeared nose heavy. Significant back pressure was required to rotate the aircraft and lift off from the runway. The aircraft exceeded the calculated take-off safety speed by about 25 kt. The aircraft rose at a higher initial climb speed than usual, but the crew did not receive any warnings. No further issues were experienced during the flight.

**Role of Data** A 'name template' was completed by a travel agent on behalf of the school group. This group was travelling from Perth to Canberra and returning back to Perth. Despite being marked as mandatory, the "Gender Description" field in this template was left blank; options for this field were "Adult", "Child" and "Infant".

As per company procedures, two days before the Perth-Canberra leg of their journey this group was 'advance accepted' into the booking system. Since the fields recording the number of children and young passengers in the group were blank, the Customer Service Agent assumed all of the group were adults. No loading-related issues were experienced during this flight.

Two days before the return flight the group was again "advance accepted" as all adults. They were checked in at Canberra Airport and assigned seats at the rear of the aircraft. The load discrepancy caused the issues noted above.

Fortunately, there were no serious consequences. However, this incident demonstrates, once again, the importance of data. This includes checking mandatory fields are completed (i.e. the completeness Data Property). The incident also highlights the potential dangers of inappropriate default data and the importance of taking opportunities to verify data as it progresses through a system.

## Sources

- Loading Issue Involving a Boeing 737, VH-VZO, http://www.atsb.gov.au/publications/investigation_reports/2014/aair/ao-2014-088.aspx, accessed on 18 November 2016.

## A.6 LOT Flight 282

**Summary** Just after take-off from Runway 09R at London Heathrow Airport (LHR), the pilots noticed that most of the information on both of the Electronic Attitude Director Indicators (EADI) and Electronic Horizontal Situation Indicators (EHSI) had disappeared. The aircraft entered Instrument Meteorological Conditions (IMC) at about 1,500 feet Above Aerodrome Level (AAL), and the co-pilot had no option but to fly using the standby attitude indicator and standby compass. He experienced difficulty in following radar headings. The aircraft returned to land at LHR after a flight of 27 minutes.

**Role of Data** The single error made by the co-pilot during the pre-flight preparation initiated the subsequent problems. This was the use of 'E' instead of 'W' when the longitude co-ordinates were entered into the Flight Management System (FMS).

The airports around London, because of their proximity to the Prime Meridian, can lead flight crews to make such co-ordinate entry errors of this nature. It is of note that the operator's route network is such that there are few destinations to the west of the Prime Meridian and hence the majority of longitude co-ordinates that need to be entered would be eastings. Inertial Reference System (IRS) alignment warnings should have alerted the crew but may have been dismissed.

This incident relates to the accuracy of the data entered into the FMS.

## Sources

- Air Accidents Investigation Branch, Bulletin 6/2008.

## A.7 Comair Flight 5191

**Summary** On 27th August 2006 Comair flight 5191 crashed during take-off from Blue Grass Airport, Lexington, Kentucky. The flight crew was instructed to take-off from runway 22, but instead lined up on runway 26 and began the take-off roll. The airplane ran off the end of the runway and impacted the airport perimeter fence, trees, and terrain. The captain, flight attendant and 47 passengers were killed.

The National Transportation Safety Board determined that the probable cause of the accident was the flight crewmembers' failure to use available cues and aids to identify the airplane's location on the airport surface during taxi and their failure to cross-check and verify that the airplane was on the correct runway before take-off.

**Role of Data** The Airport Charts used by the crew were inaccurate. The airport was under construction, and the charts were not kept current with the rapid changes that were taking place during the construction work. The chart did not accurately reflect either the taxiway identifiers and or the taxiway that was closed on the day of the accident.

Due to a previously unrecognised software glitch, any information the chart provider received after normal work hours on Fridays was not included in their regular updates. Furthermore, the chart provider modified the Blue Grass Airport chart after the accident to include a note that Runway 8/26 is "daytime VMC use only", even though this information had been published since 2001. Additionally there was a local Notice to Airmen (NOTAM) issued advising of taxiway closures due to construction work. However the crew was not provided with this information in their dispatch paperwork.

Given the preceding discussion, it is apparent that the timeliness Data Property relates to this incident.

Sources

- Attempted Takeoff from Wrong Runway - Comair Flight 5191 - Accident Report, National Transportation Safety Board, NTSB/AAR-07/05.

- Wikipedia, http://en.wikipedia.org/wiki/Comair_Flight_5191, accessed on 18 November 2016.

## A.8 RN Submarine, Trawler

**Summary** On 15 April 2015, a dived Royal Navy submarine snagged the fishing gear of the UK registered trawler *Karen*, 15 miles south-east of Ardglass, Northern Ireland. *Karen* had been trawling for prawns on a westerly heading at 2.8 knots when its fishing gear was snagged and it was dragged backwards at about 7 knots. *Karen*'s crew managed to release both winch brakes, freeing the trawl warps; the starboard warp ran out completely but the port warp became fouled on the winch drum, causing the vessel to heel heavily to port and its stern to be pulled underwater. *Karen* broke free from the submarine when the port warp parted; there was structural damage to the vessel but it returned to Ardglass safely under its own power. Evidence of the collision on board the submarine was either not observed or misinterpreted.

**Role of Data** The nature of sub-surface dived operations requires the use of Sonar technology to detect collision hazards. Detection in this way is reliant on noise emanating from contacts. In this instance the fishing trawler was detected but misidentified as a merchant vessel rather than a fishing vessel because the submarine's sonar operators did not detect or report hearing trawl noise. Given the number of vessels operating in the area, it is almost certain that the noise levels being generated would have been extremely high, with noise from one vessel masking the noise from another. Such a situation would make it very difficult for the sonar operators to methodically identify and analyse each contact, in particular to identify discrete acoustic classification clues such as trawl noise. As a result the trawler was assessed to be that of a small merchant vessel and the command team's perception would have been that no risk of collision could exist between a submarine at safe depth and a merchant vessel.

Review concluded that the submarine was operating near to the limit of its capability. Given that all the submarine's systems were reported to be functioning properly, it was clearly apparent that the submarine's limit of capability had, in reality, been exceeded, with its sonar and command teams becoming cognitively overloaded, leading to degraded situational awareness and poor decision-making.

In conclusion, the Maritime Accident Investigation Board (MAIB) report stated, "The collision happened because the submarine's command team believed *Karen* to be a merchant ship, so they did not perceive any risk of collision or need for avoiding action."

Being unable to distinguish between a trawler and a merchant vessel is an example of a failure to maintain the required resolution Data Property.

Sources

- MAIB Accident Investigation Report, https://www.gov.uk/maib-reports/collision-between-the-stern-trawler-karen-and-a-dived-royal-navy-submarine, accessed on 18 November 2016.

## A.9 Grounding of Navigator Scorpio

**Summary** On 3 January 2014, the liquefied gas carrier NAVIGATOR SCORPIO ran aground on Haisborough Sand in the North Sea. The vessel was undamaged, no pollution occurred and after two and a half hours the vessel refloated on the rising tide.

The schedule for the NAVIGATOR SCORPIO was changed close to the time of its departure. This change meant that additional North Sea coastal charts were required. These charts were delivered to the vessel shortly before its departure; they were not up to date with the latest corrections and they were not corrected prior to sailing. In addition, the passage plan (i.e., vessel route) was not checked by the master before sailing.

When the master checked the passage plan, which had been drawn up by the second officer (2/O), he suggested a change to a portion of the route. After discussion with the 2/O the route was left unchanged, but with a requirement that position fixes be obtained every five minutes rather than every fifteen. Whilst acting as the sole bridge watchkeeper the 2/O was distracted by further passage planning activities and lost positional awareness. This led to the grounding of the vessel. After the grounding false information was added to the navigation chart to give the appearance that five minute positional fixes had been taken.

**Role of Data** There was clearly an issue with the timeliness of the additional North Sea charts; these were not as up to date as was required. Likewise, there was also an issue with the timeliness of the master's check of the passage plan.

In addition, the fluidity of the chart data allowed the 2/O to make false post-grounding additions to create an incorrect impression. According to the Marine Accident Investigation Branch's report, such actions are not uncommon. These actions affect the verifiability of the chart data, which makes post-accident investigations more complicated.

**Sources**

- Grounding of the liquefied gas carrier Navigator Scorpio on Haisborough Sand, North Sea 3 January 2014, Marine Accident Investigation Branch, Report No 30/2014, November 2014.

## A.10 MS Oliva

**Summary** At about 0510 (UTC) on 16 March 2011, OLIVA, a Maltese registered bulk carrier ran aground on the north-west coast of Nightingale Island in the Tristan Da Cunha Group. OLIVA was on a loaded passage from Santos, Brazil to China. The vessel sustained severe bottom damage to almost all of her water ballast tanks that resulted in the vessel developing a 12 degree list to port.

On 18 March, the vessel broke up in two sections; the forward section drifted away and the aft section capsized and sank. All this resulted in widespread pollution around the islands of Nightingale and Inaccessible because of the diesel and fuel oil that escaped from the vessel's fuel tanks.

**Role of Data** Both the second mate and chief mate were not aware that OLIVA was heading towards Nightingale Island. This was because there was apparently no indication on the plotting chart to alert them of the dangers ahead. It appeared that the bridge team was focused on following the GPS track (red course line) superimposed on the radar screen instead of monitoring the vessel's position in relation to surrounding hazards.

'No Go' areas were not marked on the chart. It appeared that the vessel did not have BA Chart 1769, which was the appropriate large scale chart covering the Tristan Islands.

This incident highlights the importance of data resolution and availability.

Sources

- 'Safety Investigation into the grounding of the bulk carrier OLIVA On Nightingale Island, Tristan Da Cunha on 16 March 2011', Transport Malta - Marine Safety Investigation Unit, Marine Safety Investigation Report No. 14/2012.

## A.11 Sichem Osprey

**Summary** On 10 February 2010 at 0436 (local), the chemical tanker SICHEM OSPREY, on her way from Panama to Ulsan (South Korea) stranded at more than 16 knots on the north-easterly part of Clipperton Island, although an Officer Of the Watch and a lookout were on the bridge and no damage was reported prior to the accident. A 100 metre fore part of the vessel had been grounded. No pollution was observed.

**Role of Data** Anti-collision radar alarm thresholds were apparently not set according to the Captain's instructions. The adjustments were not reappraised by any of the Officers or the Captain. There were sizeable discrepancies between the fixes plotted on the chart and those displayed on the radar.

This incident highlights the role of adaptation type data, used to set radar alarm thresholds. It also relates to the Data Properties of accuracy and traceability (e.g. for the adaptation data).

Sources

- Stranding of the chemical tanker vessel SICHEM OSPREY on 10 February 2010 on Clipperton Island, Bureau d'enquêtes sur les évènements de mer.

## A.12 The Pride of Canterbury

**Summary** On 31 January 2008, the roll-on roll-off Passenger ferry, PRIDE OF CANTEBURY grounded on a charted wreck while sheltering from heavy weather in an area known as 'The Downs' off Deal, Kent. The vessel suffered severe damage to her port propeller system but was able to proceed unaided to Dover, where she berthed with the assistance of two tugs.

The vessel had been in the area for over 4 hours when, while approaching a turn at the northern extremity, the bridge team became distracted by a fire alarm and a number of telephone calls for information of a non-navigational nature. The vessel overshot the northern limit of the identified safe area before the turn was started. The Officer Of the Watch (OOW) became aware that the vessel was passing close to a charted shoal, but he was unaware that there was a charted wreck on the shoal. The officer was navigating by eye and with reference to an electronic chart system which was sited prominently at the front of the bridge, but he was untrained in the use and limitations of the system. The wreck would not have been displayed on the electronic chart due to the user settings in use at the time. A paper chart was available, but positions had only been plotted on it sporadically and it was not referred to at the crucial time.

**Role of Data** Although the Voyage Management System (VMS) was loaded with Electronic Navigational Charts (ENC) for the vessel's area of operation, the system had not been approved by the Maritime and Coastguard Agency (MCA) as the owner's policy was for the VMS to be used as an aid to navigation only, with PRIDE OF CANTEBURY's paper charts being utilised as the primary means for navigation. Relevant admiralty charts were supplied to the vessel for this purpose.

Although the VMS was not approved for use as the primary means of navigation, the officers on PRIDE OF CANTEBURY were apparently using it as if it was, despite the fact that many of them, including the Chief Officer, who was in charge at the time of the accident, were not fully trained in its use.

Among other things, this incident reflects shortcomings in the completeness Data Property of the justification and instructional data types.

Sources

- Report on the investigation into the grounding of Pride of Canterbury 'The Downs'- off Deal, Kent 31 January 2008, Marine Accident Investigation Branch, Report No 2/2009, January 2009.

## A.13 Dallas Hospital Ebola Incident

**Summary** On 26th September 2014, a Dallas hospital mistakenly sent home a man who had the Ebola virus having missed what would have appeared to be an obvious potential case: a Liberian citizen with fever and abdominal pain who said he had recently travelled from Liberia. He returned to the hospital, was eventually diagnosed with the illness, but subsequently died. Two nurses that had treated the man also contracted the virus but later recovered.

**Role of Data** There have been mixed reports on the cause of the problem, but what is clear is that external social phenomena such as the Ebola outbreak, which are outside the hospital's electronic health record (EHR) system and processes, can change the safety significance of data held in the EHR. If the importance of the data is not recognised and elevated appropriately in the support tools and processes, then the risk of unintended harm can increase.

This conclusion is reinforced by system vendors who are now updating their systems to reflect the Ebola crisis in light of the Dallas incident.

Sources

- NBC News, http://www.nbcnews.com/storyline/ebola-virus-outbreak/texas-hospital-makes-changes-after-ebola-patient-turned-away-n217296, accessed on 18 November 2016.

## A.14 Advocate Lutheran Hospital

**Summary** A Chicago hospital paid $8.25 million to settle a lawsuit brought by the parents of an infant boy who died at the institution in October 2010 after a series of medical errors.

The mother gave birth to her son 4 months prematurely. She stayed by his side with her husband for the next six weeks while boy remained in the hospital's care. On 15 October, the baby suddenly died after coming out of a heart operation without any clear complications.

**Role of Data** The hospital determined that a pharmacy technician had entered information incorrectly when processing an electronic intravenous (IV) order for the baby. This resulted in an automated machine preparing an IV solution containing a massive overdose of sodium chloride - more than 60 times the amount ordered. The problem would have been identified by automated alerts in the IV compounding machine, but those were not activated when the customised bag was prepared for the baby. That is, adaptation data had been used to change the behaviour of the machine.

Investigations also found that the outermost label on the IV bag administered to the baby did not reflect its actual contents. And while a blood test on the infant had shown abnormally high sodium levels, a lab technician assumed the reading was inaccurate. This highlights a different perspective on the dangers of defaulting (in this case a default assumption rather than a numerical default value).

Since the incident, staff have been activating alerts for similar IV compounders used in the system's hospitals and strengthened "double check" policies for all medications leaving pharmacies, among other measures.

### Sources

- Chicago Tribune, http://articles.chicagotribune.com/2012-04-05/news/chi-parents-awarded-825-million-in-infants-death-20120405_1_clear-complications-lab-technician-double-check-policies, accessed on 18 November 2016.

- Chicago Tribune, http://allhealthcare.monster.com/benefits/articles/4770-babys-death-spotlights-safety-risks-linked-to-computerized-hospital-systems, accessed on 18 November 2016.

## A.15 Cedars Sinai Medical Centre - CT Scanner

**Summary** A software misconfiguration in a Computed Tomography (CT) scanner used for brain perfusion scanning at Cedar Sinai Medical Center in Los Angeles, California, resulted in 206 patients receiving radiation doses approximately 8 times higher than intended. This error persisted for an 18 month period, starting in February 2008. Some patients reported temporary hair loss and erythema.

**Role of Data** The problem reportedly resulted from an error made by the hospital in resetting the CT machine after it began using a new protocol for the procedure in February 2008, but it wasn't detected until one of the patients reported patchy hair loss in August 2009.

"There was a misunderstanding about an embedded default setting applied by the machine," according to a statement from Cedars-Sinai. "As a result, the use of this protocol resulted in a higher than expected amount of radiation."

This incident reflects the importance of data verifiability, especially with regards to default (and adaptation) data.

### Sources

- Los Angeles Times, http://articles.latimes.com/2009/oct/10/local/me-cedars-sinai10, accessed on 18 November 2016.

- HealthImaging, http://www.healthimaging.com/topics/diagnostic-imaging/update-cedars-sinai-explains-ct-perfusion-radiation-overexposure, accessed on 18 November 2016.

## A.16 Lake Peigneur Drilling Accident

**Summary** Lake Peigneur is located in Louisiana, United States of America. It was a ten-foot deep freshwater lake popular with sportsmen. On 20th November 1980 an exploration rig drilling for oil in the lakebed was evacuated as it began to sink; this was perceived by the crew as a structural collapse. Meanwhile, the nearby Jefferson Island salt mine was being evacuated due to the sudden onset of flooding.

The rig crew had been drilling a test well into deposits alongside a salt dome under Lake Peigneur. By some miscalculation, the assembly drilled into the third level of the nearby Diamond Crystal Salt Mine. Fresh water from the lake soon began trickling into the salt mine. Over the course of the morning, the fresh lake water began dissolving the salt and enlarging the hole until water was literally flooding into the mine.

The whirlpool created as the lake drained into the mine sucked in the drilling platform, eleven barges, trees and soil. The Delcambre Canal, which usually drains from the lake into a bay on the Gulf of Mexico, had its flow reversed. This resulted in Lake Peigneur becoming a saltwater lake. Fortunately, no injuries or loss of human life were reported.

**Role of Data** Federal experts from the Mine Safety and Health Administration were not able to determine the cause of the accident due to confusion over whether the rig was drilling in the wrong place or whether the mine's maps were inaccurate. However, the incident demonstrates the potentially significant effects of either a data error, or the misinterpretation of data.

### Sources

- Wikipedia, http://en.wikipedia.org/wiki/Lake_Peigneur, accessed on 18 November 2016.

- Oil Rig Disasters, http://home.versatel.nl/the_sims/rig/lakepeigneur.htm, accessed on 18 November 2016.

## A.17 Mars Climate Orbiter

**Summary** The Mars Climate Orbiter was a spacecraft launched aboard a Delta II rocket by NASA from Cape Canaveral on 11th December 1998. Its intended mission was to study the Martian atmosphere and climate, whilst acting as a communications relay for other spacecraft on or near Mars.

The plan was that the rocket would place the spacecraft into a transfer orbit to Mars, which would be optimised along the way by a series of four trajectory correction manoeuvres. Insertion into Mars orbit was to take place at an altitude of 226km, but during the week after the final correction manoeuvre, calculations predicted that it would be between 150km and 170km; revised to 110km the day before insertion. The orbiter was able to survive atmospheric stresses down to about 80km.

On 23rd December 1999, the spacecraft passed behind Mars, and so out of radio contact, earlier than expected; communications were never regained.

Final calculations placed the spacecraft in a trajectory that would have taken it within 57km of the Martian surface, but it is likely to have disintegrated before getting to that point.

**Role of Data** It transpires that the orbiter's Flight Management System (FMS) software was designed to work with metric Newton seconds, whereas a FMS data-file generated by ground system software used pound-force seconds. A Newton is about 22.5% of a pound-force or a factor of 4.45. (See Section 4 of reference [29]).

The cost of the mission was stated by NASA to have been $327.6 million in total ($193.1 million to develop the spacecraft, $91.7 million for launch and $42.8 million for mission operations).

This incident shows the importance of tracking Data Properties (including units) throughout the entire system.

### Sources

- Wikipedia, http://en.wikipedia.org/wiki/Mars_Climate_Orbiter, accessed on 18 November 2016.

- See also [3].

## A.18 Fort Drum Artillery Incident

**Summary** Two artillery shells were fired more than a mile off target during an Army firing exercise at Fort Drum in Northern New York in March 2002. The shells landed near a mess tent where a Battalion were having breakfast. Two soldiers were killed, 13 were injured.

**Role of Data** The initial artillery site was unsuitable so the unit had to move nearly a mile from the initial site. The unit then had trouble setting up its digital and wire communications. The movement of the unit was not taken into account when programming the firing coordinates. Also, in what was termed a 'software behavioural shortfall' the system was designed to reset the gun elevation to zero. The correct altitude for the new site was not entered into the safety calculations, and the mistakes were not captured by the data review process.

This incident relates to the integrity (and possibly the accuracy) of the firing coordinates and gun elevation data.

### Sources

- The New York Times, http://www.nytimes.com/2003/07/02/nyregion/officer-found-negligent-in-deaths-of-2-at-fort-drum.html, accessed on 18 November 2016.

- AP News Archive, http://www.apnewsarchive.com/2002/Army-Reports-on-Ft-Drum-Accident/id-539bf2ea24b8dd66009c6efee2be926c, accessed on 18 November 2016.

## A.19 Interception of Communications

**Summary** In July 2015 it was reported that a public authority was undertaking an investigation into the uploading of indecent images of children and requested details of the account connected to the IP address used to upload the images. Issues with a new upgrade of the communication provider's system resulted in the incorrect data being disclosed. Investigations revealed that a further five requests had resulted in the incorrect data being disclosed. Data was acquired in six cases that related to individuals unconnected with the investigations. In one of these cases a welfare check was delayed on a child believed to be in crisis.

**Role of Data** Under the Regulation of Investigatory Powers Act 2000, Internet Service Providers and indeed other communication service providers (e.g. mobile phone network providers) are required to provide data to investigatory bodies such as the Police. This data can be used to support criminal investigation and prosecutions and in the protection of vulnerable children and adults. The data clearly has the potential to be safety related, but there is no obligation for data providers to treat it as such. In this case the data errors (i.e. loss of integrity) led to a child being exposed to additional risk of harm.

### Sources

- IOCCO, http://www.iocco-uk.info/docs/2015%20Half-yearly%20report%20(web%20version).pdf, accessed on 18 November 2016.

# Appendix B   Organisational Data Risk Assessment

*Data is becoming the new raw material of business.*
***Craig Mundie***

Note: this questionnaire only provides an initial organisational data risk level assessment. Further work is required to establish the safety data risks in detail such as determining a Data Safety Assurance Level (DSAL) for relevant data sets.

| Organisational Data Risk (ODR) Assessment Form |
|---|
| *This form is used to determine the safety risk related to data for a particular organisation and usage.* |

*This form must be completed from the perspective of **one** of the organisations involved; typically this will be the organisation using the data or the contractor supplying the system that handles the data. This form needs to be completed for each instance / application / scope / risk profile and should consider a defined boundary for the analysis, e.g. the scope of supply for the contractor or the limit of the data user's operational responsibility. It may be useful for both contracting parties to complete the form from their respective positions to check the data risk responsibilities and apportionment.*

*It is anticipated that this form will be used during early phases of a procurement or supply and also for changes to existing supplies. It can also be used to assess existing legacy scenarios.*

*Answer the questions as they apply in the context of the scope of supply. Mark the response with the "best" fit for the given scenario. Note that not all elements have to be satisfied. For each response also add a brief justification for that particular selection as opposed to any other choice.*

*If the answer to a question is completely unknown at this stage; it is suggested that the middle value or higher is chosen and an explanation added to the justification.*

*When all the relevant questions have been answered and justified, add the scores together to give a final total and record the value in the appropriate field. Use this total to determine the final ODR level based on the stated ranges.*

*The ODR level determined may be used to determine the management regime required to mitigate the risk associated with the data.*

| | | | |
|---|---|---|---|
| Data Scenario/Context Name: | | | |
| Data Scenario/Context Description: | | | |
| Scope/Data Boundary and Perspective: | | | |
| Completed By: | | Date Completed: | |

*Answer each question using the response that forms the best match for the particular scenario. Not all statements have to be satisfied and some judgement is required; it is expected that the majority of statements in the selected response can be satisfied with some interpretation. The use of multiple criteria in each question enables a smaller and manageable set of questions to be posed to provide a holistic view of the overall risk.*

| | QUESTION 1 – SEVERITY AND PROXIMITY | | |
|---|---|---|---|
| | **How severe could an accident be that is related to the data? Could it be caused directly by the data?** *This question considers the safety consequence, the proximity and contribution of the data to the accident sequence.* | | |
| 1a | All currently foreseen uses of the data could not contribute to an accident. The data is not relied upon for safe operation. Negligible environmental impact. | 1 | ☐ |
| 1b | A possible use of data could contribute to a minor accident, but only via lengthy and indirect routes. Could lead to minor injury or temporary discomfort for 1 or 2 people. Many other people/systems are involved in checking the data. Some aspects of safe operation rely very indirectly on the data. Minor environmental impact only via indirect routes. | 2 | ☐ |
| 1c | A use of the data could lead to a significant accident resulting in minor injuries affecting several people or one serious injury. Several other people/systems are involved in checking the data. There is a dependency on the data for safe operation. Environmental impact is possible. | 4 | ☐ |
| 1d | A likely use of the data could directly lead to a serious accident resulting in serious injuries affecting a number of people, or a single death. One human or independent check is involved for all data. There is major dependency on the data for safe operation. Major environmental impact is possible. | 8 | ☐ |
| 1e | An intended use of the data could lead to an accident resulting in death for several people. The accident could be caused by the data with little chance of anything else detecting and mitigating the data issues. The accident could affect the general public or cause catastrophic environmental impact. | 16 | ☐ |
| Justification: | | | |

| | QUESTION 2 – ORGANISATIONAL AND SOCIETAL IMPACT | | |
|---|---|---|---|
| | **What would be the impact on the organisation, client or public if an accident occurred related to the data?** *This question considers the tolerability within this industry sector and the general public. How much would it affect the organisation or society? Would a claim be likely? Would it generate press interest? Would a formal investigation ensue?* | | |
| 2a | Little interest, accidents happen all the time in this sector; very high societal tolerability. Negligible chance of claims or investigations. No adverse publicity likely. | 1 | ☐ |
| 2b | Some concern from the client, but accidents happen occasionally; high societal tolerability. Small chance of claim against the organisation. Local or specialist press interest. Minor investigation or audit. | 2 | ☐ |
| 2c | Public would be concerned, accidents are rare in this sector; some societal tolerability. Significant chance of claim against the organisation. Regional press interest. Client inquiry or investigation likely. | 4 | ☐ |
| 2d | Public would be alarmed and consider the accident a result of poor practice; little societal tolerability. Claims very likely. National press or media coverage a possibility. Legal or independent inquiry may follow. | 8 | ☐ |
| 2e | Public would be outraged and consider such an accident unacceptable; almost no societal tolerability. Multiple claims/fines from regulators or courts are likely. International press or media coverage. Official and/or public enquiry possible. | 16 | ☐ |
| Justification: | | | |

| QUESTION 3 – RESPONSIBILITY | | | |
|---|---|---|---|
| **How much responsibility does this organisation have for data safety?** | | | |
| *This question considers how much legal and other responsibility and ownership the organisation has for data safety aspects within this scenario. What liabilities for consequential losses/3rd party claims does the organisation have via the contract or other means? What is the scale of the organisation's contribution to the overall scope?* | | | |
| 3a | The organisation is not responsible for any data safety aspects. No liabilities for accident claims related to the data lie with the organisation. Client or other party has accepted full data safety responsibility. The organisation is fully covered and indemnified by the client or a 3rd party. | 1 | ☐ |
| 3b | The organisation is a small part of a large consortium. It has minimal liability for data safety via the contract. It is partly covered by explicit client or 3rd party protections. All safety data is managed by subcontractors, the organisation only reviews and monitors. | 2 | ☐ |
| 3c | The organisation is a significant part of the consortium team. It has some share of the data safety responsibility. Specific data safety liabilities to the client via the contract are mentioned. There are no indemnities in the organisation's favour. All key safety data obligations are explicitly flowed down to subcontractors. | 4 | ☐ |
| 3d | The organisation is prime for a small programme or has the bulk of the data safety responsibility within a team. Specific accident-related liabilities in the contract are significant. The organisation provides some indemnities to others via the contract. Some significant data safety obligations are not flowed down to subcontractors. | 7 | ☐ |
| 3e | The organisation is priming a major programme or has total data safety responsibility. Specific accident-related liabilities in the contract are large (or unlimited). The organisation provides explicit indemnities in favour of the client/3rd parties for accidents. Safety data obligations have not been discussed or are not flowed down to subcontractors. | 12 | ☐ |
| **Justification:** | | | |

| QUESTION 4 – LEGAL AND REGULATORY FRAMEWORK | | | |
|---|---|---|---|
| **What legal and regulatory environment will this work be subject to?** | | | |
| *This question considers the legal and regulatory obligations that this work will have to conform to. How well is the legal framework defined and understood? Is there an established standards culture? Is there a regulator and certification process?* | | | |
| 4a | Well understood and tested legal framework, one jurisdiction. Highly regulated sector with one overseeing body. Well established industry guidelines and standards for safety data. Formal certification processes. | 1 | ☐ |
| 4b | Understood and established legal framework, a few related jurisdictions. Regulated sector, more than one overseeing body. Industry guidelines and standards for safety data. Some formal certification processes. | 2 | ☐ |
| 4c | Some understanding of legal position, several jurisdictions. Partially regulated sector, several possible overseeing bodies. Some industry guidelines and standards that refer to data. Informal certification processes. | 4 | ☐ |
| 4d | Complex, poorly defined legal position, multiple different jurisdictions. Largely unregulated sector with no established overseeing body. Some industry guidelines and standards that mention data. Some informal certification processes. | 6 | ☐ |
| 4e | Very complex, untested and unclear legal position, many diverse jurisdictions. Unregulated sector with no overseeing body. No industry guidelines or standards for data. No certification processes. | 10 | ☐ |
| **Justification:** | | | |

| QUESTION 5 – ORGANISATIONAL MATURITY | | |
|---|---|---|
| **How mature is this organisation regarding data safety?** *This question considers the maturity of the organisation in relation to awareness and management of the risks associated with safety data. Are staff trained, managed and resourced to enable proper handling of data safety risk?* | | |
| 5a | Explicit recognition of data as a source of safety risk. Formal and established processes and procedures in place for the identification and control of safety data. Staff trained and fully aware of safety data risks. Senior management fully aware and supportive of data safety management activities. Management of safety data risks fully supported and funded. | 1 | ☐ |
| 5b | Awareness of data as a source of safety risk. Informal processes and procedures in place for the identification and control of safety data. Staff awareness of safety data risks. Senior management awareness of data safety management issues. Good support and funding for management of safety data risks. | 2 | ☐ |
| 5c | Some awareness of data as a source of safety risk. Some ad-hoc processes and procedures in place for the identification and control of safety data. Some staff awareness of safety data risks. Some senior management awareness of data safety management issues. Some support or partial funding for management of safety data risks. | 4 | ☐ |
| 5d | Little awareness of data as a source of safety risk. Minimal processes or procedures in place for the identification and control of safety data. Little staff awareness of safety data risks. Little senior management awareness of data safety management issues. Little support or minimal funding for management of safety data risks. | 7 | ☐ |
| 5e | No recognition of data as a source of safety risk. No processes or procedures in place for the identification or control of safety data. No staff training or awareness of safety data risks. Senior management not aware or in denial of safety data risks. No support or funding for management of safety data risks. | 10 | ☐ |
| Justification: | | |

| QUESTION 6 – OWNERSHIP AND USAGE | | |
|---|---|---|
| **How widely is the data used and who by?** *This question considers how much usage and what type of users there are likely to be of the data. How complex is the data supply chain? In what geographies is it used? How many owners and interfaces are there?* | | |
| 6a | Minimal or infrequent usage. One data owner, a specialist highly trained user group. Single organisation or recipient usage only. | 1 | ☐ |
| 6b | A number of operational data users. Simple linear supply chain. More than one data owner. Specialist user or limited public access. Small scale operation. No general web access. Few user organisations or recipients. | 2 | ☐ |
| 6c | Regional usage. Some public or mainstream usage. A few supply chains. A few data owners. Some web access. Several user organisations or recipients. | 4 | ☐ |
| 6d | National usage. Public or mainstream usage. Several supply chains. Several data owners. Web access. Some or varied user organisations or recipients. | 7 | ☐ |
| 6e | International usage. Extensive public or mainstream usage. Extensive web access. Many complex supply chains. Many and diverse data owners. Many and diverse user organisations or recipients. | 12 | ☐ |
| Justification: | | |

| | QUESTION 7 – SIZE, COMPLEXITY AND NOVELTY | | |
|---|---|---|---|

**What is the scale, sophistication and complexity of the data and its manipulation?**
*This question considers the nature of the data, its lifecycle and how easy it is to detect errors in the data.*

| | | | |
|---|---|---|---|
| 7a | Simple data structures. Mature and established data storage and manipulation techniques and technologies. One or two interfaces. No timeliness aspects. No transformations. Data is easily verifiable. Data is easily traceable to original source. | 1 | ☐ |
| 7b | Varied data structures. Mainstream data storage and manipulation techniques and technologies. Several interfaces. Few timeliness aspects. Few data transformations. Data is verifiable. Data is traceable to original source. | 2 | ☐ |
| 7c | Complex with some unstructured data. Current data storage and manipulation techniques and technologies. Multiple interfaces. Some timeliness aspects. Some data transformations. Data is difficult to verify. Data is difficult to trace back to original source. | 4 | ☐ |
| 7d | Complex, varied or partially unstructured data. Novel storage and manipulation techniques and technologies. Multiple complex interfaces. Time critical. Complex data transformations. Data is very difficult to verify. Data is very difficult to trace back to original source. | 7 | ☐ |
| 7e | Highly complex, varied or unstructured data. Highly novel storage and manipulation techniques and technologies. Many and complex, ill-defined or dynamic interfaces. Highly time critical. Many and complex data transformations. Data is infeasible to verify. Data is impossible to trace back to original source. | 10 | ☐ |

Justification:

| | QUESTION 8 – BOUNDARIES AND INTERFACES | | |
|---|---|---|---|

**How well defined and understood are the boundaries and interfaces for this data scenario?**
*This question considers the number, complexity and definition status of the boundaries and interfaces where data is exchanged. How well understood are the boundaries and interfaces? Are standard formats and protocols used? Is data exchange time critical? Are all assumptions and ambiguities relating to the data exchange resolved?*

| | | | |
|---|---|---|---|
| 8a | One well-understood boundary and few, well-defined interfaces. Standard interface formats and protocols. No timeliness aspects to data exchange. No remaining ambiguities, TBCs or TBDs. No assumptions. | 1 | ☐ |
| 8b | A few, understood boundaries and several defined interfaces. Mainly standard interface formats and protocols. Few timeliness aspects to data exchange. Few areas of ambiguity, few TBCs and TBDs. Few assumptions. | 2 | ☐ |
| 8c | Several, established boundaries, some defined, some undefined and some ambiguous interfaces. Mixture of standard and non-standard interface formats and protocols. Some timely data exchanges. Some areas of ambiguity, some TBCs and TBDs. Some assumptions. | 4 | ☐ |
| 8d | Many, poorly understood boundaries, many undefined or ambiguous interfaces. Mostly non-standard interface formats and protocols. Time sensitive data exchange. Many areas of ambiguity, many TBCs and TBDs. Many assumptions. | 6 | ☐ |
| 8e | A large number of unclear boundaries; a large number of unknown and undefined interfaces. Completely non-standard, complex interface formats and protocols. Real-time data exchange. Large areas of ambiguity, a large number of TBCs and TBDs. A large number of assumptions. | 10 | ☐ |

Justification:

| ORGANISATIONAL DATA RISK LEVEL | |
|---|---|
| Record the total score and use it to determine the ODR level based on the ranges given below. <u>If the first 3 question's scores sum up to 6 or less then disregard the scores for the remaining questions.</u> | |
| Score 14 or less | ODR0 |
| Score 15 to 21 | ODR1 |
| Score 22 to 37 | ODR2 |
| Score 38 to 47 | ODR3 |
| Score 48 and above | ODR4 |
| Total Score for this scenario/context: | |
| ODR Level for this scenario/context: | |

# Appendix C   Data Safety Culture Questionnaire

> *Data is the fabric of the modern world: just like we walk down pavements, so we trace routes through data and build knowledge and products out of it.*
> **Ben Goldacre**

This form helps an organisation appreciate the data safety culture. It can be applied at various levels, including at the project level and at the organisational level.

| Data Safety Culture Questionnaire Form |
| --- |
| *This form is used to assess the safety culture related to data for a particular programme (the DSC value).* |

You play a key role in protecting the organisation from data safety risks and your views are important. This self-assessment survey is designed to assess our current level of data safety culture within the programme. The output can help us to improve our safety position.

Please tick the box which reflects your view and answer as honestly as possible. Space is provided for explanatory comments. Your response will only be of value if it reflects what you actually believe is the case, rather than what you believe should happen.

If you would like to remain anonymous please print and send this form by post.

The survey should take no longer than 10 minutes. It is anticipated that this form will be used on a regular basis (e.g. annually).

| Programme Name: | |
| --- | --- |
| Completed By: | Date Completed: |

*Answer each question as you see it - there is no right answer!*

| QUESTION 1 – MY VIEW OF OUR SUPPLY | | Don't Know | Strongly Disagree | Disagree | Maybe | Agree | Strongly Agree |
| --- | --- | --- | --- | --- | --- | --- | --- |
| 1a | I see data as an important factor in the safety of my programme. | ☐ | ☐ | ☐ | ☐ | ☐ | ☐ |
| 1b | I am familiar with the safety aspects of our data. | ☐ | ☐ | ☐ | ☐ | ☐ | ☐ |
| 1c | I understand how data in our solution can contribute to an accident. | ☐ | ☐ | ☐ | ☐ | ☐ | ☐ |
| 1d | I think we could be blamed if there were an accident due to our data. | ☐ | ☐ | ☐ | ☐ | ☐ | ☐ |
| Comments: | | | | | | | |

| QUESTION 2 – WHAT WE'RE DOING | | Don't Know | Strongly Disagree | Disagree | Maybe | Agree | Strongly Agree |
|---|---|---|---|---|---|---|---|
| 2a | I think that the programme is aware of data safety risks. | ☐ | ☐ | ☐ | ☐ | ☐ | ☐ |
| 2b | I believe we need to implement measures to manage data safety risks. | ☐ | ☐ | ☐ | ☐ | ☐ | ☐ |
| 2c | I think that the programme meets its obligations (e.g. has a Data Management Plan in place and a role with specific responsibilities in this area). | ☐ | ☐ | ☐ | ☐ | ☐ | ☐ |
| Comments: | | | | | | | |

| QUESTION 3 – MY ROLE | | Don't Know | Strongly Disagree | Disagree | Maybe | Agree | Strongly Agree |
|---|---|---|---|---|---|---|---|
| 3a | I know my role relates to the management of data and associated safety risks. | ☐ | ☐ | ☐ | ☐ | ☐ | ☐ |
| 3b | If I had a safety concern about our data I would report it. | ☐ | ☐ | ☐ | ☐ | ☐ | ☐ |
| 3c | I know who the data safety representative is on my programme. | ☐ | ☐ | ☐ | ☐ | ☐ | ☐ |
| 3d | I have received adequate training regarding data safety for my role. | ☐ | ☐ | ☐ | ☐ | ☐ | ☐ |
| 3e | I feel supported in dealing with data safety risks. | ☐ | ☐ | ☐ | ☐ | ☐ | ☐ |
| 3f | I have adequate time to address any data safety issues. | ☐ | ☐ | ☐ | ☐ | ☐ | ☐ |
| Comments: | | | | | | | |

# Appendix D   Supplier Data Maturity

> Processed data is information. Processed information is knowledge. Processed knowledge is Wisdom.
> *Ankala V. Subbarao*

This questionnaire may be used for two purposes:

1. To support a procurement process - distributed by an organisation looking for a company that can handle safety-critical development, because the system they require to be developed is known to have safety-critical requirements

2. Internal audits - used internally by a company developing systems with expanding safety-critical data which needs to assure itself of its capability to fulfil customer needs.

## D.1 Data Safety Supplier Questionnaire

### Organisation

1. For each software development involving data, is there a designated data safety manager?

2. If so, does the data safety manager report directly to the project manager?

3. Are the management reporting channels for data assurance and software development separate?

4. Is data subject to a formal configuration control process?

5. Is data engineering represented on the system design team?

6. Is data engineering process improvement part of the company quality systems?

### Resources, Personnel and Training

1. Are personnel specified as responsible for data safety as a separate role from software and system design and development?

2. Is there a required training programme for data specialists?

3. Is training on data safety issues part of the training for managers or management teams?

4. Is there a formal training for programme for data safety design and review leaders

### Data Issues Growth Management

1. Is a mechanism employed for maintaining awareness of the state of the art in data safety technology?

2. Is a mechanism employed for comparing the company approach to data safety with external processes for data safety practised elsewhere in the industry?

3. Is a mechanism used for introducing new technologies and processes into system development?

4. Is a mechanism in place for identifying and replacing obsolescent processes related to data safety?

## Documented Standards and Procedures

1. Describe any formal procedure adopted at each periodic management review of the status of data related to the system?

2. Describe any method used for ensuring that the data development team understands each data requirement.

3. Is a safety-critical mechanism used for assessing the use of existing data in new applications?

4. Are data test cases developed formally with a company standard?

5. Is there a document which describes how the customer is to be consulted over data issues?

6. Is particular care taken to capture requirements, design, review and test data for man-interfaces?

7. Is there a data risk monitoring and tracking to closure procedure practised?

## Process Metrics

1. Are statistics of failures due to data errors during development kept to feedback and learn from in future development?

2. Are data issue action items tracked to closure and reports maintained of causes?

3. Is configuration data separately developed from everyday operational data?

4. Is data test coverage measured and recorded?

5. Are all states, from which configuration data will be required, tested, (including emergency reboot), and results recorded?

6. Are analyses of errors due to data conducted to determine their process related causes?

7. Are the process causes reviewed and changes to processes implemented where appropriate?

## Process Control

1. Is regression testing routinely performed when errors are discovered?

2. Is the adequacy of regression testing subject to an assurance process to ensure new errors are not introduced?

3. Is a mechanism used for identifying and resolving system engineering issues that affect data?

4. Is a mechanism used for ensuring traceability between the data requirements and the top level design?

5. Is the importance of data in the system engineering process reviewed to maintain processes at an adequate level to cope with the expanding role of data in the Internet of Things?

# Appendix E    Data Types – Detail

> It's difficult to imagine the power that you're going to have when so many different sorts of data are available.
> **Tim Berners-Lee**

The following table provides additional information, in the form of explanations and lists of typical containers, for the identified Data Types.

| No. | Type | Description | Explanation | Typical containers |
|-----|------|-------------|-------------|--------------------|
| Context | | | | |
| 1 | Predictive | Data used to model or predict behaviours and performance | Data for studies, models, prototypes, initial risk assessments, etc. This is the data produced during the initial concept phase which subsequently flows into further development phases. | Prototype results, evaluations, analyses |
| 2 | Scope, Assumption and Context | Data used to frame the development, operations or provide context | Restrictions, risk criteria, usage scenarios, etc. explaining how the system will be used and any limitations of use. | Concepts of operation, Safety Case Report Part 1 |
| 3 | Requirements | Data used to specify what the system has to do | Data encompassing requirements, specifications, internal interface or control definitions, data formats, etc. | Formal specifications, interface control documents, user requirements documents, Safety Case Report Part 1 |
| 4 | Interface | Data used to enable interfaces between this system and other systems: for operations, initialisation or export from the system | Data that exists to enable exchange between this system and other external systems. Covers start-of-life operations (data import or migration), end-of-life operations and ongoing operational exchange of data between systems. | Protocols, schemas, interface control documents, transition plans, Extract-Transform-Load tool specifications, cleansing and filtering rules |
| 5 | Reference or Lookup | Data used across multiple systems with generic usage | Data comprising generic reference information sets used by multiple systems (i.e., not produced solely for this system). Typically updated infrequently, and not specific to this system. | Dictionaries, materials information, sector data reference sets, encyclopedias |
| Implementation | | | | |
| 6 | Design and Development | Data produced during development and implementation | Data encompassing the design and development process artefacts: everything from design models and schemas to document review records. It also includes test documents (specification and results) but not the test data itself. | Design documents, review records, hardware, software, test scripts, code inspection reports, Safety Case Report Part 2 |

| No. | Type | Description | Explanation | Typical containers |
|-----|------|-------------|-------------|-------------------|
| 7 | Software | Data that is compiled and executed to achieve the desired system behaviour | From some perspectives it is helpful to consider software (e.g., source code) as another type of data. | Text files, configuration management systems |
| 8 | Verification | Data used to test and analyse the system | Data comprising the test values and test data sets used to verify the system. It may include real data, modified real data or synthetic data. It includes data used to drive stubs, and any data files used by simulators or emulators. | Test data sets, stub data, emulator and simulator files |
| Configuration | | | | |
| 9 | Infrastructure | Data used to configure, tailor or instantiate the system itself | Data used to set up and configure the system for a particular installation, product configuration, or network environment. | Network configuration files, initialisation files, hardware pin settings, network addresses, passwords |
| 10 | Behavioural | Data used to change the functionality of the system | Data to enable / disable or configure functions or behaviour of the system. | XML configuration files, Comma Separated Variable (CSV) data, schemas |
| 11 | Adaptation | Data used to configure to a particular site | Data used to tailor or calibrate a system to a particular physical site or environment, incorporating physical or environmental conditions. | Configuration files |
| Capability | | | | |
| 12 | Staffing and Training | Data related to staff training, competency, certification and permits | Data which allows staff to perform a function within the wider context of the safety-related system. This may include training records, competency assessments, permits to work, etc. | Human Resources records, training certificates, card systems |
| The Built System | | | | |
| 13 | Asset | Data about the installed or deployed system and its parts, including maintenance data | Data related to location, condition and maintenance requirements of the system under consideration. This may cover hardware, software and data. | Inventory, asset and maintenance database systems |
| 14 | Performance | Data collected or produced about the system during trials, pre-operational phases and live operations | Data produced by and about the system during introduction to service and live service itself. Includes fault data and diagnostic data. This may be the results of various phases of introduction and may include trend analysis to look for long-term problems. | Field data, Support calls, bug reports, non-compliance reports, Defect Reporting And Corrective Action System (DRACAS) data |

| No. | Type | Description | Explanation | Typical containers |
|---|---|---|---|---|
| 15 | Release | Data used to ensure safe operations per release instance | Explanation of particular features or limitations of a release or instance. May include specific time-limited workarounds and caveats for a release. | Release notes, Certificates of Design (CoDs), Transfer documents, Safety Case Report Part 2 or Part 3 |
| 16 | Instructional | Data used to warn, train or instruct users about the system | Data that explains to users the risks of the systems and gives any mitigations that may be required to be implemented by users, e.g., by process, procedure, workarounds, limitations of use. | Manuals, Standard Operating Procedures (SOPs), on-line help, training courses, Safety Case Report Part 3 |
| 17 | Evolution | Data about changes after deployment | Data that covers enhancements, formal changes, workarounds, and maintenance issues. It also covers data produced by configuration management activities, such as baselines or branch data. | Change requests, modification requests, issue and version data, configuration management system outputs |
| 18 | End of Life | Data about how to stop, remove, replace or dispose of the system | Data covering all activities related to taking the system out of service or mothballing / storage / dormant phases. | Transition, disposal and decommissioning plans |
| 19 | Stored | Data stored by the system during operations | Data stored or utilised within the system which has end-user meaning. It may be displayed an used within the system or may be for transfer and distribution to other systems or downstream users. It is data that has some real domain meaning. | May be stored internally within the system (e.g., in databases or text files), or transferred into or out of the system through interfaces (e.g., Ethernet) |
| 20 | Dynamic | Data manipulated and processed by the system during operations | Data processed, transformed or produced by the system which has end-user meaning. It may be displayed an used within the system or may be for transfer and distribution to other systems or downstream users. It is data that has some real domain meaning. | May be manipulated within the system in data structures or transferred into or out of the system through interfaces |
| | | | Compliance and Liability | |
| 21 | Standards and Regulatory | Data that governs the approaches, processes and procedures used to develop safety systems. | Data predominantly in the form of documents that describe and dictate the activities, processes, competencies etc. to be used for a particular development in a particular sector. | Standards documents, guidelines, legal directives and laws |

| No. | Type | Description | Explanation | Typical containers |
|-----|------|-------------|-------------|-------------------|
| 22 | Justification | Data used to justify the safety position of the system | Data used to justify, explain and make the case for starting or continuing live operations and why they are safe enough. Often passed to external bodies (e.g., regulators, Health and Safety Executive, Independent Safety Auditors) for their review. | Safety Case Report, certification case, regulatory documents, COTS justification file, design justification file |
| 23 | Investigation | Data used to support accident or incident investigations (i.e., potential evidence) | Data collected or produced during an incident or accident investigation which may be used in investigation reports, lessons learnt or prosecutions. This can be process data, trace data, site data (e.g., photographs of crash site) or may be derived (accident simulations, analyses, etc.). | Incident/accident investigation reports and supporting documents |
| Meta-Property | | | | |
| +1 | Trustworthiness | (Meta) data which tells us how much the system can be trusted | Data which provides assurance or confidence about the other data within or about the system under consideration. This may be some of the data mentioned in the other types, but may be different. | Data audits, data quality index measures, sign-off sheets traceability records, model database |

# Appendix F    HAZOP Guidewords - Detail

*I don't see the logic of rejecting data just because they seem incredible.*
**Fred Hoyle**

The following table provides a more detailed description of HAZOP guidewords.

| Property | Description | HAZOP Data Properties | HAZOP Data Guidewords |
|---|---|---|---|
| Integrity | The data is correct, true and unaltered | Loss, partial loss, incorrect, multiple | Correctness, truth, original, trustworthy, coherency, stability, perfect, unquestionable, faithful, certain, ordered, unadulterated, unmodified, unchanged, clean, uncontaminated, untainted, proper, flawless, organized, exact, undistorted, faultless, guided, connected, linked, traced, unbiased. |
| Completeness | The data has nothing missing or lost | Loss, partial loss, incorrect, multiple | Whole, complete, entire, finished, done, stable, qualified, certified. |
| Consistency | The data adheres to a common world view, e.g., units | Loss, partial loss, incorrect, multiple, too early, too late, loss of sequence | Coherent, compatible, congruent, congruous, harmonious, deconflicted, consistent, appropriate, suitable, sound, cleansed. |
| Continuity | The data is continuous and regular without gaps or breaks | Loss, partial loss, incorrect, late, loss of sequence | Smooth, continuous, regular, gapless, whole, complete, entire, unfragmented. |
| Format | The data is represented in a way which is readable by those that need to use it | Loss, partial loss, incorrect, multiple | Conformant, suitable, valid, configured, well-formed, setup, composed, well structured, arranged, compliant, organised, exact, unaliased, migrated, transformed. |
| Accuracy | The data has sufficient detail for its intended use | Loss, partial loss, incorrect, multiple | Accurate, true, correct, undistorted, unbiased, faultless. |
| Resolution | The smallest difference between two adjacent values that can be represented in a data storage, display or transfer system | Loss, partial loss, incorrect, multiple | Exact, untruncated, retention of detail, clarity, determination, distinguishable, clear, within range, distinct, separated, discernible, discriminatable, unconfused, divisible, unaliased, granularity, precision. |
| Traceability | The data can be linked back to its source or derivation | Loss, partial loss, incorrect, multiple, too early, too late, loss of sequence | Traceable, verifiable, indexed, linked, connected, justified, proven, evidenced, substantiated, continuous, unfragmented, complete, networked. |
| Timeliness | The data is as up to date as required | Loss, partial loss | Timely, early, ready, expected, unique, appropriate, opportune, ordered, organised, anticipated, seasonable, converging, settling, on-time, latency, lag, lead time, time slots, real-time, determinism, predictable. |

| Property | Description | HAZOP Data Properties | HAZOP Data Guidewords |
|---|---|---|---|
| Verifiability | The data can be checked and its properties demonstrated to be correct | Loss, incorrect, partial loss, multiple, too early, too late, loss of sequence | Verifiable, provable, checkable, supportable, demonstrable, sustainable, certifiable, defensible, excusable, justifiable, undisputable, irrefutable, validated. |
| Availability | The data is accessible and usable when an authorized entity demands access | Loss, partial loss, multiple, too early, too late | Ready, available, obtainable, reachable, accessible, serviceable, operable, functional, usable, capable, released, issued, disseminated, distributed. |
| Fidelity / Representation | How well the data maps to the real world entity it is trying to model | Loss, incorrect, partial loss, multiple, too early, too late | Representative, accurate, faithful, trustworthy, characteristic, normal, standard, real, expected, natural, typical, regular, fit for purpose, validated, separable, associated, correct units/dimensions, stable, unbiased. |
| Priority | The data is presented / transmitted / made available in the order required | Loss, incorrect, partial loss, multiple, too early, too late | Current, ordered, included, precedence, hierarchy, pre-eminence, retained, ahead, readiness. |
| Sequencing | The data is preserved in the order required | Loss, incorrect, partial loss, multiple | Ordered, contiguous, unique, ordered, clear, continuous, successive, uninterrupted, sequential. |
| Intended Destination / Usage | The data is only sent to those that should have them | Loss, incorrect, partial loss, multiple, too early, too late, loss of sequence | Directed, delivered, copied, sent, transmitted, correct recipient, unintercepted, unseen, integral, received, acknowledged, forwarded, filtered. |
| Accessibility | The data is visible only to those that should see them | Loss, incorrect, partial loss, multiple, too early, too late | Secure, open, visible, reachable, seen, usable, accessible, obtainable, uncompromised, secure, encrypted, preserved. |
| Suppression | The data is intended never to be used again | Loss, incorrect, partial, too early, too late, too much, too little | Hidden, encrypted, private, confidential, erased, unlinked, unavailable, unaccessible, redacted. |
| History | The data has an audit trail of changes | Loss, incorrect, partial loss, multiple | Justifiable, traceable, provable, supportable, demonstrable, sustainable, certifiable, defensible, excusable, justifiable, undisputable, irrefutable. |
| Lifetime | When does the safety-related data expire | Loss, too early, too late, incorrect, multiple, loss of sequence | Expiry date, age, validity, currency, applicability, durability, duration, lifespan, stretch, tenure, half-life, longevity, span, in-date, best-before, window, established. |
| Disposability / Deletability | The data can be permanently removed when required | Loss, incorrect, partial, too early, too late | Unavailable, unaccessible, redacted, hidden, filtered, lost, deleted, destroyed, backup, archive, locked, secured, unlinked. |

# Appendix G   Data Safety Management Plan

> *Things get done only if the data we gather can inform and inspire those in a position to make a difference.*
> **Mike Schmoker**

This section gives a suggested Data Safety Management Plan (DSMP) table of contents. It is expected that this will be needed only for aspects not already covered in a Safety Management Plan (SMP), or similar. It can be merged with an SMP, if appropriate. However it may be useful to consider the distinct data perspective by using a DSMP as well as an SMP. Regardless, a close connection should be maintained between the SMP and the DSMP.

Data Safety Management Plan suggested contents:

1.  Introduction:

    –  Scope and Context (Sets the scene, describes the project, scenario, concept of operations, etc.);

    –  Boundaries and Interfaces (Describes the main interfaces and exchanges, with a scope boundary diagram.);

    –  Owners (Who owns the data under consideration as it progresses through the system?);

    –  Producers / Consumers (Who are the producers and consumers of the data the system inputs and outputs?);

    –  Assumptions;

    –  References; and

    –  Abbreviations and Acronyms.

2.  Analysis of Assigned DSAL and ODR Level (Implications of the data analyses.):

    –  System Integrity Level (SIL), etc., Implications (What impact does the DSAL have on the required SIL, or similar measure, and vice versa?);

    –  Development Implications (Are there any special development considerations? Derived from the SIL if there is one, otherwise what is deemed necessary for this system.);

    –  Verification Implications (Derived from the SIL if there is one, otherwise what is deemed necessary for this system.);

    –  Assurance Implications (Derived from the SIL if there is one, otherwise what is deemed necessary for this system.); and

    –  Process / Procedure Implications (Derived from the SIL if there is one, otherwise what is deemed necessary for this system.).

3.  Types of Safety Data in Scope (A list of all the types to be considered in the system context.).

4.  Data Requirements Analysis:

    –  Lifecycles (What data lifecycles are to be used?);

– Specific Targets (Are there any qualitative or quantitative targets for the data?); and

– Security Considerations (How will security be managed in this context? Are there any security / safety conflicts? Are there any security-related causes of data hazards?).

5. Management Approach (How will the organisation manage the data safety risks?):

   – Organisation;

   – Responsibilities;

   – Authorisations; and

   – Approvals and Signoffs.

6. Justification Approach (How will the safe usage of the data be justified, e.g., as part of the Safety Case Report?).

7. Analyses / Verifications to be Performed (What analyses or checks are to be performed on the data?).

8. Documents to be Produced (The list of documents to be produced related to data aspects.).

9. Appendix: DSAL Guidelines Response (Tailored version of the tables from this document. What is considered applicable / useful and what is not?).

# Appendix H  Acronyms, Definitions & Glossary

*The plural of anecdote is not data.*
**Mark Bekoff**

## H.1 Acronyms

| Acronym | Expansion |
| --- | --- |
| AAL | Above Aerodrome Level |
| ARP | Aerospace Recommended Practice |
| ARQ | Automatic Repeat-reQuest |
| ATM | Air Traffic Management |
| ATSB | Australian Transport Safety Bureau |
| BIT(E) | Built-In Test (Equipment) |
| CBT | Computer Based Technologies |
| CNS | Communications, Navigation, Surveillance |
| CoD | Certificate of Design |
| COTS | Commercial Off-The-Shelf |
| CRC | Cyclic Redundancy Check |
| CSV | Comma Separated Variable |
| CT | Computed Tomography |
| DA | Data Artefact |
| DBMS | Database Management System |
| DME | Distance Measuring Equipment |
| DRACAS | Defect Reporting And Corrective Action System |
| DSAL | Data Safety Assurance Level |
| DSG | Data Safety Guidance |
| DSIWG | Data Safety Initiative Working Group |
| DSMP | Data Safety Management Plan |
| DVD | Digital Versatile Disc |
| EADI | Electronic Attitude Direction Indicators |
| ECS | Electronic Chart System |
| ECU | Electronic Control Unit |
| EHSI | Electronic Horizontal Situation Indicators |
| FMGS | Flight Management Guidance System |
| FMS | Flight Management System |
| FDAL | Function Development Assurance Level |
| GP | General Practitioner |
| HAZOP | Hazard and Operability Study |
| HSE | Health and Safety Executive |
| ICD | Interface Control Document |
| ICAO | International Civil Aviation Organisation |
| IDAL | Item Development Assurance Level |

| Acronym | Expansion |
|---------|-----------|
| IMC | Instrument Meteorological Conditions |
| ISA | Independent Safety Auditor |
| ISO | International Organisation for Standardisation |
| LHR | London Heathrow |
| NaN | Not a Number |
| NOTAM | Notice to Airmen |
| ODR | Organisational Data Risk |
| OEM | Original Equipment Manufacturer |
| OOW | Officer Of the Watch |
| OSI | Open Systems Interconnection |
| SAP | Service Access Point |
| SCADA | Supervisory Control And Data Acquisition |
| SCCI | Standardisation Committee for Care Information |
| SCSC | Safety Critical Systems Club |
| SDU | Service Delivery Unit |
| SIL | Safety Integrity Level |
| SMP | Safety Management Plan |
| SOP | Standard Operating Procedure |
| SoS | Systems of Systems |
| SSS | Safety-critical Systems Symposium |
| USB | Universal Serial Bus |
| VMS | Voyage Management System |
| VOR | VHF Omnidirectional Range |
| WOW | Weight On Wheels |
| XML | eXtensible Markup Language |

## H.2 Definitions & Glossary

| | Definition | Source |
|---|---|---|
| **A** | | |
| Accuracy | Closeness of agreement between a test result and the accepted reference value. Note that a test result can be observations or measurements. | ISO 19113:2005 [19] |
| | A degree of conformance between the estimated or measured value and the true value. | (EU) No 73/2010 [13] |
| Accuracy (temporal) | Correctness of the temporal references of an item (reporting of error in time measurement). Correctness of ordered events or sequences, if reported. Validity of data with respect to time. | ISO 19138:2006 [22] |
| (data) Assurance Level | The required assurance level for the aeronautical data process is identified, based on the overall system architecture through allocation of risk determined using a preliminary system safety assessment. | RTCA/DO-200A [10] |

| | Definition | Source |
|---|---|---|
| | An indication of how much assurance is required (commensurate to risk) before deploying software into an operational system. | J Spriggs, based on (EC) No 482/2008 [6] |
| Adaptation Data | Data used to customise elements of the Air Traffic Management System for their designated purpose. Adaptation data is utilised to customize elements of the CNS / ATM system for its designated purpose at a specific location. These systems are often configured to accommodate site-specific characteristics. These site dependencies are developed into sets of adaptation data. Adaptation data includes data that configures the software for a given geographical site, and data that configures a workstation to the preferences and / or functions of an operator. Examples include, but are not limited to: a) Geographical Data - latitude and longitude of a radar site. b) Environmental Data - operator selectable data to provide their specific preferences. c) Airspace Data - sector-specific data. d) Procedures - operational customisation to provide the desired operational role. Adaptation data may take the form of changes to either database parameters or take the form of pre-programmed options. In some cases, adaptation data involves re-linking the code to include different libraries. Note that this should not be confused with recompilation in which a completely new version of the code is generated. | ED-153 [8] |
| Aeronautical Data | A representation of aeronautical facts, concepts or instructions in a formalised manner suitable for communication, interpretation or processing. | (EU) No 73/2010 [13] |
| | Data used for aeronautical applications such as navigation, flight planning, flight simulators, terrain awareness, and other purposes. | RTCA/DO-178C [7] |
| Application Data | Data used in the system during operations: this is the data processed or produced by the system which has end-user meaning. It may be displayed and used within the system or may be for transfer or distribution to other systems or downstream users. It is data that has some real "application" meaning, i.e., is not to do with the system internals. | SCSC Data Safety Initiative Working Group |
| Assumption Data | Data used to frame the development, operations or provide context: restrictions, risk criteria, usage scenarios, etc. explaining how the system will be used and any limitations of use. | SCSC Data Safety Initiative Working Group |
| Availability | The property of being accessible and usable upon demand by an authorized entity. | ISO27001:2005 [25] |
| **B** | | |
| **C** | | |
| Completeness | Completeness of the data provided. | RTCA/DO-200A [10] |
| Configuration Data | Data that configures a generic software system to a particular instance of its use. | (EC) No 482/2008 [6] |
| | Data used to configure, tailor or instantiate the system: data used to set up and configure the system to perform a particular function, for a particular installation, product configuration, behaviour or specific usage. | SCSC Data Safety Initiative Working Group |

| | Definition | Source |
|---|---|---|
| | Data that configures a generic software system to a particular instance of its use (e.g., data for flight data processing system for a particular airspace, by setting the positions of airways, reporting points, navigation aids, airports and other elements important to air navigation). | ED-153 [8] |
| Confidentiality | The property that information is not made available or disclosed to unauthorized individuals, entities, or processes. | ISO27001:2005 [25] |
| (data) Correctness | Completeness, self consistency, protection against alteration or corruption and consistency with the functional requirements of the data driven system. | IEC 61508 Part 3 [26] |
| (data) Coupling | The dependence of a software component on data not exclusively under the control of that software component. | RTCA/DO-178C [7] |
| (data) Criticality | Classification of data by the potential effect of erroneous data on the expected operation that is supported by that data. | RTCA/DO-200A [10] |
| Critical Data | Data with an integrity level as defined in Chapter 3, Section 3.2 point 3.2.8(a) of Annex 15 to the Chicago Convention, i.e., integrity level one in one hundred million: there is a high probability when using corrupted critical data that the continued safe flight and landing of an aircraft would be severely at risk with the potential for catastrophe. | (EU) No 73/2010 [13] |
| Customisation (data) | Data used to configure a system or component. | Def(Aust)5679 [4] |
| **D** | | |
| Data | A thing given or granted; something known or assumed as fact, and made the basis of reasoning or calculation; an assumption or premiss from which inferences are drawn. | Oxford English Dictionary (OED) |
| | A reinterpretable representation of information in a formalized manner suitable for communication, interpretation or processing. | ISO/IEC 2382 [24] |
| Database | A set of data, part or the whole of another set of data, consisting of at least one file that is sufficient for a given purpose or for a given data processing system. | RTCA/DO-178C [7] |
| Data Chain | An 'Aeronautical Data Chain' is a conceptual representation of the path that a set, or element of aeronautical data takes from its creation to its end use. An aeronautical data chain is a series of interrelated links wherein each link provides a function that facilitates the origination, transmission and use of aeronautical data for a specific purpose. | RTCA/DO-200A [10] |
| | A collection of organisational data processing functions, where data is transferred from one chain participant to another between data origination and end use. | P. Ensor [11] |
| | Any combination of two or more data elements, data items, data codes, and data abbreviations in a prescribed sequence to yield meaningful information; for example, "date" consists of data elements year, month, and day. | McGraw-Hill Dictionary [28] |
| (data) Dictionary | The detailed description of data, parameters, variables, and constants used by the system. | RTCA/DO-178C [7] |
| Data Driven Systems | System which relies upon configuration data or lookup tables to define the functionality of the system. | IEC 61508 Part 4 [27] |
| Data Intensive System | Systems which make extensive use of large amounts of data. | N. Storey [32] |

| | Definition | Source |
|---|---|---|
| Design & Development Data | Data produced during development and implementation: this is data encompassing the design & development process artefacts: everything from design models and schemas to document review records. It also includes test documents (specification and results) but not the test data itself. | SCSC Data Safety Initiative Working Group |
| **E** | | |
| End of Life Data | Data about how to stop, remove, replace or dispose of the system: this is data covering all activities related to taking the system out of service or mothballing / storage / dormant phases. | SCSC Data Safety Initiative Working Group |
| (data) Error | Discrepancy with the universe of discourse. | ISO 19138:2006 [22] |
| | Discrepancy between a data value and the true, specified or theoretically correct value or condition. | P. Ensor [11] |
| Essential Data | Data with an integrity level as defined in Chapter 3, Section 3.2 point 3.2.8(b) of Annex 15 to the Chicago Convention, i.e., integrity level one in one hundred thousand: there is a low probability when using corrupted essential data that the continued safe flight and landing of an aircraft would be severely at risk with the potential for catastrophe. | (EU) No 73/2010 [13] |
| Evolution Data | Data about changes after deployment, i.e., data that cover enhancements, formal changes, workarounds, and maintenance issues. It also covers data produced by configuration management activities, such as baselines or branch data. | SCSC Data Safety Initiative Working Group |
| **F** | | |
| **G** | | |
| **H** | | |
| (data) Hazard | Use of data in the context of a system that could lead to an accident. | SCSC Data Safety Initiative Working Group |
| **I** | | |
| Information | Knowledge communicated concerning some particular fact, subject, or event; that of which one is apprised or told - intelligence, news - as contrasted with data. | Oxford English Dictionary (OED) |
| | Knowledge that has a contextual meaning. | ISO/IEC 2382 [24] |
| Information (aeronautical) | Information resulting from the assembly, analysis and formatting of aeronautical data. | (EU) No 73/2010 [13] |
| (data) Integrity | The assurance that a data element retrieved from a storage system has not been corrupted or altered in any ways since the original data entry or latest authorised amendment. | RTCA/DO-200A [10] |
| | The degree of assurance that a data item and its value have not been lost or altered since the data origination or authorised amendment. | (EU) No 73/2010 [13] |
| | The degree of undetected (at system level) non-conformity of the input value of the data item with its output value. | (EU) No 1207/2011 [12] |
| | The property of protecting the accuracy and completeness of assets, i.e., that which has value to the organisation. | ISO27001:2005 [25] |

| | Definition | Source |
|---|---|---|
| Instructional Data | Data used to warn, train or instruct users about the system: this is data that explains to users the risks of the systems and gives any mitigations that may be required to be implemented by users, e.g., by process, procedure, workarounds, limitations of use. | SCSC Data Safety Initiative Working Group |
| (data) Item | Single attribute of a complete data set, which is allocated a value that defines its current status. | (EU) No 73/2010 [13] |
| Interface Data | Data used to enable interfaces between systems: for operations, initialisation or export from the system: data that exists to enable exchange between systems. Covers start-of-life operations (data import or migration), end-of-life operations and ongoing operational exchange of data between systems. | SCSC Data Safety Initiative Working Group |
| Investigation Data | Data to support accident or incident investigations (i.e. potential evidence: this is data collected or produced during an accident investigation which may be used in investigation reports, lessons learnt or prosecutions. This can be source data (e.g., photographs of crash site) or may be derived (accident simulations, analyses, etc.). | SCSC Data Safety Initiative Working Group |
| **J** | | |
| Justification Data | Data used to justify the safety position of the system: data used to justify, explain and make the case for starting or continuing live operations and why they are safe enough. Often passed to external bodies (regulators, HSE, ISAs) for their review. | SCSC Data Safety Initiative Working Group |
| **K** | | |
| **L** | | |
| **M** | | |
| Meta-data | Data that represents information about data itself. Note that one should distinguish between "Structural Meta-data", which is data about the design and specification of data structures (and is more properly called "data about the containers of data") and "Descriptive Meta-data", which is about individual instances of application data, the data content. | J. Inge [18] |
| **N** | | |
| **O** | | |
| Operational Data | Data collected or produced about the system during trials, pre-operational phases and live operations: data produced by and about the system during introduction to service and live service itself. Includes fault data and diagnostic data. This may be the results of various phases of introduction and may include trend analysis to look for long-term problems. | SCSC Data Safety Initiative Working Group |
| (data) Origination | Creation of a new data item with its associated value, the modification of the value of an existing data item or the deletion of an existing data item. | (EU) No 73/2010 [13] |
| **P** | | |
| (data) Product | Dataset or dataset series that conforms to a data product specification. | BS EN ISO 19131:2008 [21] |
| Predictive Data | Data used to model or predict behaviours and performance: Data for studies, models, prototypes, initial risk assessments, etc. This is the data produced during the initial concept phase which subsequently flows into further development phases. | SCSC Data Safety Initiative Working Group |

| | Definition | Source |
|---|---|---|
| **Q** | | |
| (data) Quality | A degree or level of confidence that the data provided meet the requirements of the user. These requirements include levels of accuracy, resolution, assurance level, traceability, timeliness, completeness, and format. | RTCA/DO-200A [10] |
| | Process by which the Electronic Chart Systems (ECS) Database is produced, the source materials, the resolution and reproduction accuracy of chart features, and the correctness and completeness of data. | ISO 19379:2003 [23] |
| | A degree or level of confidence that the data provided meets the requirements of the data user in terms of accuracy, resolution and integrity | (EU) No 73/2010 [13] |
| (data) Quality Attributes | Accuracy, resolution, assurance level, traceability, timeliness, completeness and format. | RTCA/DO-200A [10] |
| **R** | | |
| Release Data | Data used to ensure safe operations per release instance: explanation of particular features or limitations of a release or instance. May include specific time-limited workarounds and caveats for a release. | SCSC Data Safety Initiative Working Group |
| Requirements Data | Data used to specify what the system has to do: data encompassing requirements, specifications, internal interface or control definitions, data formats, etc. | SCSC Data Safety Initiative Working Group |
| Resolution | The smallest difference between two adjacent values that can be represented in a data storage, display or transfer system. | RTCA/DO-200A [10] |
| | A number of units or digits to which a measured or calculated value is expressed and used. | (EU) No 73/2010 [13] |
| Routine Data | Data with an integrity level as defined in Chapter 3, Section 3.2 point 3.2.8(b) of Annex 15 to the Chicago Convention, i.e., integrity level one in one thousand: there is a very low probability when using corrupted routine data that the continued safe flight and landing of an aircraft would be severely at risk with the potential for catastrophe. | (EU) No 73/2010 [13] |
| **S** | | |
| (data) Set | Identifiable collection of data. Note that a dataset may be a smaller grouping of data which, though limited by some constraint such as spatial extent or feature type, is located physically within a larger dataset. Theoretically, a dataset may be as small as a single feature or feature attribute contained within a larger dataset. A hardcopy map or chart may be considered a dataset. | BS EN ISO 19131:2008 [21] |
| Software Lifecycle Data | Data that is produced during the software lifecycle to plan, direct, explain, define, record, or provide evidence of activities (including the software product itself). This data enables the software lifecycle processes, system or equipment approval and post-approval modification of the software product. | ED-153 [8] |
| Staffing and Training Data | Data related to staff training, competency, certification and permits: data which allows staff to perform a function within the wider context of the safety-related system. This may include training records, competency assessments, permits to work, etc. | SCSC Data Safety Initiative Working Group |

| | Definition | Source |
|---|---|---|
| Standards and Regulatory Data | Data that governs the approaches, processes and procedures used to develop safety-related systems: this is data predominantly in the form of documents that describe and dictate the activities, processes, competencies etc. to be used for a particular development in a particular sector. | SCSC Data Safety Initiative Working Group |
| System Data | Data about the installed or deployed system and its parts, including maintenance data: data related to location, condition and maintenance requirements of the system under consideration. This may cover hardware, software and data. | SCSC Data Safety Initiative Working Group |
| **T** | | |
| Timeliness | A measure of the time delay between a change in the real world and the associated database update being available to the user. | P. Ensor [11] |
| | The difference between the time of output of a data item and the time of applicability of that data item. | (EU) No 1207/2011 [12] |
| Traceability | Ability to determine the origin of the data. | RTCA/DO-200A [10] |
| Trace (data) | Data providing evidence of traceability of development and verification processes software lifecycle data without implying the production of any particular artifact.Trace data may show linkages, for example, through the use of naming conventions or through the use of references or pointers either embedded in or external to the software lifecycle data. | RTCA/DO-178C [7] |
| **U** | | |
| **V** | | |
| (data) Validation | The activity whereby a data element is checked as having a value that is fully applicable to the identity given to the data element, or a set of data elements that is checked as being acceptable for their purpose. | RTCA/DO-200A [10] |
| | Process of ensuring that data meets the requirements for the specified application or intended use. | (EU) No 73/2010 [13] |
| Validity (period of) | Period between the date and time on which aeronautical information is published and the date and time on which the information ceases to be effective. | (EU) No 73/2010 [13] |
| (data) Verification | Evaluation of the output of an aeronautical data process to ensure correctness and consistency with respect to the inputs and applicable data standards, rules and conventions used in that process. | (EU) No 73/2010 [13] |
| Verification Data | Data used to test and analyse the system: this is data comprising the test values and test data sets used to verify the system. It may include real data, modified real data or synthetic data. It includes data used to drive stubs, and any data files used by simulators or emulators. | SCSC Data Safety Initiative Working Group |
| **W** | | |
| **X** | | |
| **Y** | | |
| **Z** | | |

# Appendix I    References

*The goal is to turn data into information, and information into insight.*
**Carly Fiorina**

[1]    American Airlines Flight 965, http://en.wikipedia.org/wiki/American_Airlines_Flight_965

[2]    Disastercast Episode 18 Friendly Fire [including Data Safety], Drew Rae, http://disastercast.co.uk/transcripts/episode-18-transcript/

[3]    Disastercast Episode 15 Quantitative Nonsense [including Mars Climate Orbiter], Drew Rae, http://disastercast.co.uk/transcripts/episode-15-transcript/

[4]    DEF(AUST)5679, Issue 2, Safety Engineering for Defence Systems - Standard, October 2008

[5]    Die Lage der IT-Sicherheit in Deutschland 2014, Bundesamt fuer Sicherheit in der Informationstechnik, December 2014, https://www.bsi.bund.de/SharedDocs/Downloads/DE/BSI/Publikationen/Lageberichte/Lagebericht2014.pdf

[6]    Commission Regulation (EC) No 482/2008 of 30 May 2008 establishing a software safety assurance system, as amended, http://eur-lex.europa.eu/LexUriServ/LexUriServ.do?uri=OJ:L:2008:141:0005:0010:EN:PDF

[7]    RTCA/DO-178C, EUROCAE Document ED-12C, Software Considerations in Airborne Systems and Equipment Certification, January 2012

[8]    EUROCAE Document ED-153, Guidelines for ANS Software Safety Assurance - use for definitions only

[9]    RTCA/DO-330, EUROCAE Document ED-215, Software Tool Qualification Considerations, January 2012

[10]    RTCA/DO-200A, EUROCAE Document ED-76, Standards for Processing Aeronautical Data, September 1998

[11]    Safety Analysis of Navigational Data, Paul Ensor, September 2009

[12]    Commission Implementing Regulation (EU) No 1207/2011 of 22 November 2011 laying down requirements for the performance and the interoperability of surveillance for the single European sky, http://eur-lex.europa.eu/LexUriServ/LexUriServ.do?uri=OJ:L:2011:305:0035:0052:EN:PDF

[13]    Commission Regulation (EU) No 73/2010 of 26 January 2010 laying down requirements on the quality of aeronautical data and aeronautical information for the single European sky, http://eur-lex.europa.eu/LexUriServ/LexUriServ.do?uri=OJ:L:2010:023:0006:0027:EN:PDF

[14]    FAA TC-14/49, March 2015, Selection of Cyclic Redundancy Code and Checksum Algorithms to Ensure Critical Data Integrity, https://www.tc.faa.gov/its/worldpac/techrpt/tc14-49.pdf

[15]    Data Integrity - an often-ignored aspect of safety systems, Alastair Faulkner, 2004, (EngD thesis), http://wrap.warwick.ac.uk/1212/

[16]     An Assessment Framework for Data-Centric Systems, A. Faulkner, M. Nicholson. In Addressing Systems Safety Challenges, Proceedings of the Twenty-second Safety-Critical Systems Symposium, Bristol, UK. Edited by Chris Dale and Tom Anderson, ISBN 978-1491263648

[17]     The Principles of Software Safety Assurance, R. Hawkins, I. Habli, T. Kelly, 31st International System Safety Conference, Boston, Massachusetts USA, 2013

[18]     Improving the Analysis of Data in Safety-Related Systems, James Inge, 12 September 2008, http://www.safety.inge.org.uk/20080912-Inge2008b_Improving_the_Analysis_of_Data_in_Safety_Related_Systems-U.pdf

[19]     BS EN ISO 19113:2005, Geographic Information. Quality Principles

[20]     ISO31000:2009, Risk Management - Principles and Guidelines. First Edition, 2009-11-15

[21]     BS EN ISO 19131:2008, Geographic Information. Data Product Specifications

[22]     ISO/TS 19138:2006, Geographic Information. Data Quality Measures

[23]     BS ISO 19379:2003, Ships and Marine Technology. ECS databases. Content, Quality, Updating and Testing

[24]     ISO/IEC 2382-1:1993, Information Technology. Vocabulary. Part 1: Fundamental Terms

[25]     BS ISO/IEC 27001:2005, Information Technology. Security Techniques. Information Security Management Systems. Requirements

[26]     BS EN 61508-3:2010, Functional safety of electrical/electronic/ programmable electronic safety-related systems. Software Requirements, June 2010

[27]     BS EN 61508-4:2010, Functional safety of electrical/electronic/ programmable electronic safety related systems. Definitions and Abbreviations, June 2010

[28]     McGraw-Hill Dictionary of Scientific and Technical Terms, 6th Edition, November 2002. ISBN 978-0070423138

[29]     Mars Climate Orbiter Mishap Investigation Board Phase I Report November 10, 1999, http://sunnyday.mit.edu/accidents/MCO_report.pdf

[30]     Writing Secure Code, David LeBlanc, Michael Howard, Mircosoft Press, December 2002. ISBN 978-0735617223

[31]     SAE Aerospace Recommended Practice 4754A, Guidelines for Development of Civil Aircraft and Systems, December 2010

[32]     The Characteristics of Data in Data-intensive Safety-related Systems, Neil Storey & Alastair Faulkner. Lecture Notes in Computer Science, Volume 2788, 396-409, 2003

# Appendix J    DSIWG History

The task of developing generally applicable, pan-sector guidance for data safety issues was taken on by the Data Safety Initiative Working Group (DSIWG) of the Safety Critical Systems Club. The DSIWG's work started with a seminar *"How to Stop Data Causing Harm"*, which was held in December 2012; material from this seminar is available at: http://scsc.org.uk/e209 (accessed 8 December 2016). The first meeting of the group agreed the following vision:

**To have clear guidance on how data (as distinct from software and hardware) should be managed in a safety-related context, which will reflect emerging best practice.**

The group, comprising industry, academics, government and independent consultants, produced an initial guidance document in January 2014. A subsequent version of the document was released in January 2015, with a second seminar *"How to Stop Data Causing Harm: What You Need to Know"* held in December 2015; material from this seminar is available at: http://scsc.org.uk/e343 (accessed 8 December 2016). To help disseminate information, members of the DSIWG have presented papers relating to data safety in a number of fora, including the Safety-critical Systems Symposium (SSS).

A further revision of the guidance document, which included new material generated during and as a consequence of DSIWG meetings, was issued in January 2016. This latest version, issued in January 2017, has incorporated the latest thinking in this developing area.

# Appendix K    Contributors

*Without data, you're just another person with an opinion.*
**W. Edwards Deming**

This document has had the benefit of contributions from a large number of people, who work for a variety of organisations, which collectively span a range of different sectors. Note that contributions have been made on an individual basis and, in particular, the inclusion of an organisation in the following list does **not** necessarily mean that organisation agrees with the entire contents of the document.

Significant contributors to the document include:

- Mike Ainsworth, Ricardo
- Rob Ashmore, Dstl
- Michael Aspaturian, EDF Energy
- Divya Atkins, Mission Critical Applications
- Janette Baldwin, Thales UK
- Dave Banham, Rolls-Royce plc
- Ian Bingham
- John Bragg, MBDA UK Ltd
- Eric Bridgstock, Raytheon UK
- Simon Brown, QinetiQ
- Dermot Martin Burke, BAE Systems
- Dale Callicott, DKCSC Ltd
- John Carter, General Dynamics
- Martyn Clarke
- Steve Clugston, TSC
- Dijesh Das, AMEC / BAE Systems
- Duncan Dowling, DARD
- Andrew Eaton, CAA
- Ashraf El-Shanawany, CRA Risk Analysis
- Paul Ensor, Boeing
- Alastair Faulkner, Abbeymeade
- Derek Fowler, JDF Consultancy
- Ken Frazer, KAF

- Paulo Giuliani, EDF Energy
- Ian Glazebrook, Atkins
- Rob Green, NATS
- Nick Hales, DE&S
- Amira Hamilton, CGI UK and Cranfield University
- Paul Hampton, CGI UK
- Louise Harney, PA Consulting
- Ali Hessami, Vega Systems
- David Higgins
- Pete Hutchison, RPS
- Gavin Jones
- Tim Kelly, University of York
- Andrew Kent, CGI UK
- Brent Kimberley, Durham, Canada
- Julian Lockett, Frazer-Nash Consultancy Ltd
- David Lund, David Lund Consultants
- Dave Lunn, Thales UK
- Nasser Al Malki, University of York
- Victor Malysz, Rolls-Royce
- John McDermid, University of York
- Mark Nicholson, University of York
- Yvonne Oakshott, Agusta Westland
- Robert Oates, Rolls-Royce plc
- Mike Parsons, NATS
- David Perrin, Virtual PV
- Andrew Rankine
- Felix Redmill, SCSC
- Sam Robinson, EDF Energy
- Tim Rowe, EC Harris
- Alan Simpson, Ebeni

- Dave Smith, Frazer-Nash Consultancy Ltd

- John Spriggs, NATS

- Carolyn Stockton, BAE Systems

- Mark Templeton, QinetiQ

- Lesley Winsborrow

- Fan Ye, ESC

# Appendix L   Acknowledgements

> *Our ability to do great things with data will make a real difference in every aspect of our lives.*
> ***Jennifer Pahlka***

The document contributors would like to thank:

- The Safety Critical Systems Club (SCSC).

- Mike Parsons for chairing the working group meetings.

- Nick Hales of MOD DES DTECH for significant support and encouragement.

- Brian Jepson of the SCSC for web hosting support and technical help with the SCSC web site.

- Paul Hampton, Louise Harney, Lesley Winsborrow and Rob Ashmore for taking notes at meetings.

- Chris Tapp (Keylevel Consultants Limited) for his assistance in the production of this document.

- All the organisations hosting working group meetings, including:

    - Boeing;

    - Civil Aviation Authority;

    - CGI;

    - Ebeni;

    - EDF Energy;

    - Frazer-Nash Consultancy;

    - Ministry of Defence;

    - NATS;

    - QinetiQ;

    - Raytheon UK;

    - Rolls-Royce plc;

    - Thales;

    - UK Hydrographic Office;

    - University of Nottingham;

    - University of Westminster; and

    - University of York.

- All the organisations that have provided support to the document's contributors.

- Those that have been unable to attend meetings but have made supporting contributions.